D1450507

BOWLING GREEN STATE UNIVERSITY

DISCARDED

LIBRARY

THE TELEVISION AUDIENCE:
PATTERNS OF VIEWING

The Television Audience: Patterns of Viewing

An Update

Second Edition

G.J. GOODHARDT
*City University
Business School*

A.S.C. EHRENBERG
London Business School

M.A. COLLINS
*Social and Community
Planning Research*

*and
Aske Research Ltd*

Gower

JEROME LIBRARY-BOWLING GREEN STATE UNIVERSITY

© Aske Publications Ltd 1975, 1987

All rights reserved. No part of this publication may be reproduced, stored in a retrieval system, or transmitted in any form or by any means, electronic, mechanical, photocopying, recording, or otherwise without the prior permission of Gower Publishing Company Limited.

Published by
Gower Publishing Company Limited
Gower House
Croft Road
Aldershot
Hants GU11 3HR
England

Gower Publishing Company
Old Post Road
Brookfield
Vermont 05036
USA

British Library Cataloguing in Publication Data
The television audience: patterns of viewing.—2nd ed.
  1. Television audiences—Great Britain
  I. Goodhardt, G.J.
  302.2'345   HE8700.66.G7

Library of Congress Cataloging-in-Publication Data
The Television audience.

  Rev. ed. of: The television audience/G.J. Goodhardt,
A.S.C. Ehrenberg, M.A. Collins. 1975.
  Bibliography: p.
  Includes index.
  1. Television audiences—Great Britain. I. Goodhardt,
G.J. (Gerald Joseph), 1930–   . II. Ehrenberg, A.S.C.
  Television audience.
HE8700.66.G7T44 1986   302.2'345'0941   86-4658

ISBN 0 566 05083 8

Typeset by Activity Ltd., Salisbury, Wilts.
Printed in Great Britain by Blackmore Press, Shaftesbury, Dorset

Contents

The Authors   vi

Preface   vii

Preface to the First Edition   ix

Foreword to the First Edition   xi

Detailed Contents   xiii

1	TELEVISION PROGRAMMING	1
2	AUDIENCE FLOW	8
3	VIEWING DIFFERENT PROGRAMMES	19
4	OTHER FACTORS IN AUDIENCE OVERLAP	35
5	REPEAT-VIEWING	51
6	TOTAL TV VIEWING	70
7	AUDIENCE FLOW IN THE US	77
8	AUDIENCE APPRECIATION	87
9	THE LIKING OF PROGRAMME TYPES	100
10	TELEVISION AS A MEDIUM	114

IBA Reports   125

References   128

Index   131

The Authors

G.J. Goodhardt has spent 20 years in business, mostly in marketing research and advertising. In January 1975 he took up his first academic position as Reader in Marketing at Thames Polytechnic and in 1981 was appointed to the Sir John E. Cohen Chair of Consumer Studies at City University Business School.

A.S.C. Ehrenberg has spent 15 years in industry and has also held academic positions at the Universities of Cambridge, Columbia, Durham, London, Pittsburgh and Warwick. He was appointed Professor of Marketing at the London Business School in 1970 and WCM Professor and Director of the Centre for Marketing and Communication at the LBS in 1984.

M.A. Collins has worked in survey research since 1962. He is Director of the Survey Methods Centre, Social and Community Planning Research and Visiting Professor in Market Research at the City University Business School.

All three authors have been Chairman of the Market Research Society, have published widely and are consultants to Aske Research Ltd.

Preface

There has been a continuing demand for *The Television Audience* since it was first published in 1975. The message of our new edition is that the same patterns of viewing still hold, despite the passage of time and changes in the media scene. For example, we originally noted in 1967 that repeat-viewing of regular programmes in the UK had been only about 55%. Analyses of UK and US data in 1985 show that it is generally still only about 55%, or less.

We have updated some of the numerical results and illustrations in this edition, have been able to simplify some of the wording and have incorporated some new results, mainly in Chapters 6 and 8. But we have not attempted to incorporate all the additional findings about the television audience since 1975 – that would require a different book.

For new readers, this book still presents the main patterns of TV audience behaviour that are known: for those who have read it before, the main message is that the earlier results still apply.

The structure of the book

Television viewing is greatly influenced by the programmes that are on offer. In Chapter 1 we therefore briefly outline the nature of television programming and of audience measurement procedures.

Chapter 2 introduces the main concepts of audience flow – the extent to which different programmes, and different episodes of the same programme, have viewers in common. The basic 'duplication of viewing law' and 'repeat-viewing' concepts are introduced here. Chapters 3 and 4 (given new titles in this edition) develop more detailed aspects of the audience overlap between the audiences of different programmes, the question of channel loyalty, the effects of time of day and day of week and the 'inheritance' of the audience for successive programmes.

Chapter 5 examines repeat-viewing, i.e. how many viewers watch successive episodes of the same programme, and how this builds up for a longer series of episodes. Chapter 6 on the total amount of people's viewing in a week incorporates new material on how they distribute this across different channels and programme types.

Chapter 7 extends the results in the UK to viewing behaviour in the USA. Although some of the conditions are different, the main results apply, as also checked in more recent analyses.

Chapters 8 and 9 move from viewers' *behaviour*, i.e. their choice of programmes to view, to their expressed appreciation or liking of the programmes. Audience appreciation is examined, especially how it relates to the size of the audiences and the percentage of repeat-viewers.

Chapter 10 draws together the various findings and briefly sets out some implications of our analyses for the nature of television as a medium.

The methods of TV audience measurement are nowadays more widely known and are in any case changing with technological developments and we have therefore curtailed their discussion in this edition. Details in different countries can usually be obtained from the local audience measurement companies.

Acknowledgements for the updated edition

Most of the information about the UK in this book comes from studies carried out by Aske Research Ltd for the IBA (as listed at the end of the book). We have also been able to make use of recent work on US television carried out with our colleague Patrick Barwise at the London Business School and supported by the John and Mary R. Markle Foundation of New York. We are grateful to IBA, Arbitron and Markle for permission to quote reports prepared for them and to JICTAR, BARB, AGB, TAM, and Leo Burnett to quote from their data.

GJG
ASCE
MAC
October 1985

Preface to the First Edition (Extracts)

The average family in the United Kingdom currently watches television for more than five hours a day. Individuals on average watch three hours or so a day. In the great majority of homes the television set is therefore switched on for most of the evening. A similar pattern occurs in the United States and much of Western Europe.

With television occupying such a significant part of the leisure time of many people, it is not surprising that the medium has become a subject of major social and political concern and even controversy. Much of the discussion has centred on the likely effects of television. More operational questions concern the number of different TV channels, how they should be run and so on. Various studies on both sides of the Atlantic and elsewhere have tried to deal with these and similar problems.

Many of these studies have been somewhat disappointing, being piecemeal in their analysis and unconvincing in the generality of their conclusions. One possible reason has been that the studies usually failed to deal with perhaps the most basic question of all: the sort of television viewing that people actually do. Before we can learn about the effects of particular types of television on people, we need to take into account what they watch.

Implicit in certain criticisms of programmes on crime and violence is for example the thought that such programmes tend to attract the same type of regular viewer or 'addict', on whom they then have a harmful effect. As will be shown in this book, it is easy to explore the first premise – whether programmes of a given type in fact attract a particular group of people. Until this is determined, the effects of such programmes can hardly be realistically studied.

The main purpose of this book is in fact to describe how people view. We examine their loyalty to particular programmes or types of programmes, their loyalty to particular television channels and the nature of their switching between channels and programmes.

We also examine data on the audience's appreciation of programmes. Most

ix

people feel that it is not enough to assess television programmes solely in terms of the size of the audiences they attract (the *ratings*). Aside from questions of quality versus quantity, audience size is an inadequate measure of a programme's inherent appeal because ratings are affected by the programmes on alternative channels, by the views of different family members, by time of day and so on.

Most of the findings in this book are simple. Patterns are found which tend to be regular and hence become predictable. Such regularities can therefore become a basis for understanding the medium, for making forecasts about the nature of the audiences for different programmes or programme schedules and for testing ideas about the impact of different forms of television policy. Since our past work has lain more in the study of viewing patterns than in the wider issues of policy, the book itself concentrates on the audience's viewing behaviour as such.

Television is a subject on which many people have views and preconceptions. On perusing an earlier account of one of the findings here, one lay-reader concluded that the report seemed to be saying:

A person who has just been watching a detective-type programme on one channel will tend to switch to another channel if that is then showing another programme of the same type.

On being told the report actually showed the opposite effect (i.e. that there was *no* such special tendency), the reader replied: 'That is what I actually thought it said, but I couldn't believe it.' Trying to correct widely held misconceptions is one practical value of the results of systematic research.

Foreword to the First Edition

Where facts are few, myths abound. There can be few better areas for illustrating this observation than television. Despite the collection of enormous masses of data on a continuous basis over many years by both broadcasting organisations in the United Kingdom, systematic studies of how the viewer actually *behaves* – the pattern of his or her viewing – have hitherto been remarkably scarce. Professor Ehrenberg and his colleagues have taken a first but important step towards rectifying this state of affairs.

This book has arisen out of a programme of research undertaken for the Independent Broadcasting Authority by Goodhardt, Ehrenberg and Collins of Aske Research Ltd, London, which started in 1967 and is still continuing. The basic aim throughout has been to get beyond the detailed and specific information supplied by audience measurement studies, such as those by Television Audience Measurement Ltd (TAM) and Audits of Great Britain Ltd (AGB) for JICTAR (and BARB more recently), to more generalised findings about viewer behaviour. The fact that the data are derived from the same panel of viewers each day makes it possible to do so and in the process various assumptions that have been apt to be taken for granted are subjected to critical scrutiny.

Are there identifiable groups of viewers who express common programme preferences and who actually exhibit similar viewing behaviour patterns? Are viewers demonstrably 'loyal' to a channel and if so does this vary from ITV to BBC1 and BBC2? Can the audience be 'caught early' by clever arrangement of the schedule and then 'held throughout the evening'? What proportion of the audience to a programme on a given channel will see a different programme on another day? Will it be greater if the second programme is on the same day at the same time a week later, or if it is linked in some way, as a complementary programme, as part of a series, or as another episode in a serial?

All who have been concerned with television have heard confident and seemingly authoritative assertions about these and related questions which are typical of the subject matter covered in *The Television Audience*. These are important questions because they are the very stuff from which programme, scheduling and broadcasting policy decisions are created. It is probable that the implications of some of the findings in the book will come as a surprise to its readers: it is hoped that they will be helpful in the future, when such matters are discussed, as they certainly will be.

The Television Audience contains material which will be of interest not only to broadcasters but also to those who, while outside the area of active broadcasting, are concerned with the media in general and the social issues associated with mass communication. Research workers will also find it instructive as a demonstration of making fuller use of data, extensively and expensively acquired and then not always studied as deeply as they should be.

The authors have made a significant contribution to knowledge in an important area and at a time when the issues which are raised are matters of public concern and argument. I commend this book to the thoughtful consideration of the reader.

Ian R. Haldane
Head of Research
Independent Broadcasting Authority
1975

Detailed Contents

1	TELEVISION PROGRAMMING	1
+1.1	Historical Background	1
1.2	The New Distribution Channels	3
1.3	The Programmes Available	4
+1.4	Audience Measurement	6
1.5	Summary	7
2	AUDIENCE FLOW	8
2.1	Characterising the Audience	8
2.2	Interpretative Norms: High, Low or Normal?	9
2.3	Audience Duplication	9
2.4	Repeat-Viewing	13
2.5	Ratings Levels	15
2.6	Some Statistical Considerations	15
2.7	Summary	16
	Appendix: The Duplication Law	17
3	VIEWING DIFFERENT PROGRAMMES	19
3.1	The Effect of the Channels	19
3.2	Two ITV Programmes	20
3.3	Other ITV Programmes	21
3.4	The Duplication of Viewing Law	22
3.5	Viewing on Non-Consecutive Days	24
3.6	Channel Switching	25
3.7	Duplication on BBC1	27
3.8	Duplication on BBC2	27
3.9	Switching to BBC2	29
3.10	The Overall Pattern	31
3.11	The Patterns over Time	32
3.12	Summary	34

4   OTHER FACTORS IN AUDIENCE OVERLAP   35
4.1   Demographic Factors   35
4.2   Viewing at Off-Peak Times   36
4.3   Week-end Viewing   38
4.4   The Inheritance Effect   40
4.5   Programme Types   42
4.6   A Specific Application: News Broadcasts   45
4.7   Summary   50

5   REPEAT-VIEWING   51
5.1   'Z-Cars'   51
5.2   55% Repeat-Viewing   53
5.3   Viewing Other Channels   57
5.4   Non-Consecutive Episodes   58
5.5   Repeat-Viewing and Low Rating Levels   59
5.6   Audience Cumulation   62
5.7   Theoretical Models   64
5.8   Summary   67
      Appendix: The Beta-Binomial Model   68

6   TOTAL TV VIEWING   70
6.1   Hours Viewed   70
6.2   The Two Major Channels   71
6.3   The Two Smaller Channels   71
6.4   Multi-Channel Viewing   72
6.5   Viewing of Programme Types   74
6.6   Lighter Viewers of Television   75
6.7   Summary   76

7   AUDIENCE FLOW IN THE US   77
7.1   Repeat-Viewing of Different Episodes in the Same Week 77
7.2   Duplication between Different Programmes   81
7.3   The Inheritance Effect   84
7.4   Summary   86

8   AUDIENCE APPRECIATION   87
8.1   The Demand for Measures of Appreciation   87
8.2   The Appreciation Index   88
8.3   Appreciation Index and Audience Size   89
8.4   Appreciation Index and Repeat-Viewing   92
8.5   Appreciation Index and Individual Repeat-Viewing   93

8.6 'Coronation Street': A Case History 96
8.7 Interesting or Enjoyable? 97
8.8 Summary 98

9 THE LIKING OF PROGRAMME TYPES 100
9.1 Programme Clusters 100
9.2 Programme Character 107
9.3 The Nature of Programme Clusters 108
9.4 Reconciliation with Viewing Behaviour 110
9.5 Summary 113

10 TELEVISION AS A MEDIUM 114
10.1 Simple Findings 114
10.2 Some Broad Implications 116
10.3 The Pull of the Box 119
10.4 The Effects of Television 120
10.5 Summary 124

IBA Reports 125

References 128

Index 131

1 Television Programming

Every year millions in money and man-hours are poured into television – its technology, programmes and advertisements. The object of all this effort is to reach and communicate with the audience – to entertain it, stimulate it, inform it, educate it or influence it in some way.

Yet though television has become an integral part of modern life there is still no satisfactory way to evaluate the success or failure of these efforts. How much satisfaction do viewers derive from television? What needs or wants, however defined, remain unfulfilled? How great is television's influence on the audience and on society in general? Is it advantageous or detrimental?

The answers to these and other questions matter. But the questions can hardly begin to be answered until one knows how the audience in fact behaves – the patterns of people's viewing. Hence this book – a general picture of the regularities that exist in the behaviour of viewers and in their attitudes towards programmes. The aim is to contribute to a better understanding of the medium and to demolish some of the shibboleths that have found their way into popular belief.

Our topic here is relatively narrow – the television audience as such, rather than the broader interpretative issues of the role of television in society. Most of the studies that form the basis of this book were carried out by us in the UK and some in the US. We therefore concentrate here largely on British viewing patterns. But the general approach and many of the fundamental results will apply also in other countries, as is illustrated with data for the US in Chapters 2 and 7.

1.1 Historical background

The United Kingdom: over-air TV

Television transmissions began on a regular basis in the UK in November 1936. By the outbreak of World War II there were about 25,000 sets in use; now there is virtual saturation with some 20 million.

The UK had only one channel until 1955. This monopoly was held by the

1

BBC (the British Broadcasting Corporation), a public corporation financed from revenues derived by issuing annual licences for sets to the public.

In 1954 the government established another public corporation, now called the Independent Broadcasting Authority or IBA. This currently appoints 15 regional independent television (ITV) programme companies in 14 areas to provide broadcasts on a second channel. ITV is financed entirely by advertising revenue.

In 1964 parliament awarded another channel to the BBC. The intention was that BBC1 should aim at majority interest groups and BBC2 at various special interests. In 1982 Channel 4 was established as a subsidiary company of the IBA, supported by advertising, with the aim of broadcasting programmes which were 'different'.

Since January 1968 colour transmissions have been developed and used on all channels in the UK.

The amount of advertising material that can be shown on ITV and Channel 4 in the UK is controlled by law with a maximum daily average of 6 minutes per hour. The use of advertising differs in other countries (e.g. Pragnell 1985). In Germany, for example, advertising is concentrated into one limited time-band starting at 7 pm and television is funded by this with a licence fee.

The United States: over-air TV

The effective start of TV transmissions in the US was delayed by commercial competition involving differences in broadcasting techniques and types of receivers used. But in 1941 the Federal Communications Commission (FCC) – a federal agency that awards broadcasting licences and oversees the broadcasting media – authorised commercial operation on 18 VHF channels; television broadcasting was limited to six experimental stations. Further delay followed in 1945 over the merits of monochrome and VHF channels versus colour and UHF. An FCC verdict decided in favour of the former in 1947 and the American TV gold rush was on.

By 1948 there were 41 stations serving 23 cities and important advertisers had begun experimenting with the medium and sponsoring programming. By 1950, 10 million sets had been sold and by 1968 this number had risen to 90 million. Now an increasing number of homes have more than one set.

There are over 200 separate 'markets' covered by their own separate over-air TV stations. In some of the largest, like New York and Los Angeles, people can receive six or eight or more over-air stations.

Up to a quarter or so of all TV stations are non-commercial, made up mainly of the Public Broadcasting System (PBS) and various community stations. All of these tend to attract small audiences.

The 700 commercial stations are financed by advertising revenue. In most markets three stations are affiliated to the three national network companies CBS, NBC and ABC whose programmes are generally broadcast simultaneously across the country (but bearing in mind that different time-zones exist). They tend to dominate the ratings and are highly competitive towards each other. In addition there are some 20 regional networks and various independent commercial stations.

There is no law or FCC ruling governing the amount of advertising that may be carried on US television stations but the National Association of Broadcasters has a voluntary code that specifies a maximum of 16 minutes per hour.

1.2 The new distribution channels

Since the first edition of this book was published there has been something of a revolution in the distribution channels used for television:

– *Cable* (especially wide-spread in the USA, largely for reception reasons, and in Switzerland, Belgium and the Netherlands but more problematic elsewhere)
– *Satellite* (especially used to feed cable in the States, but as DBS – direct broadcast to the home – more problematic)
– *Video Cassette Recorders* (high penetration particularly in the UK – almost 40% in 1984 – with marked growth in the US only just taking off).

The scene is still changing (including some retractions of over-enthusiastic cable expansion and DBS plans). In the US, cable has led to a noticeable decline in the total share of the three networks by 10% or more in the mid-eighties. In the UK the use of VCRs is on average of the order of 10% of total viewing in VCR homes (e.g. 2 or 3 hours a week) of which much is time-shift of recorded over-air broadcasts, the rest being viewing of rented rather than bought tapes.

The revolution, however, goes less deep than had been expected by many. Most of the programmes seen are still essentially the same (with new feature films typically being seen two or three years *earlier* on pay-TV channels than before). These developments have been and will be widely discussed in the press and elsewhere and will not be described further here.

Detailed information on viewing patterns is as yet thin and likely to change in its specifics. But the overwhelming impression is that people's basic use of television – the kinds of results about repeat-viewing and programme choice described in this book – have not changed radically and, we believe, may well

<u>not change</u> in the future. They seem to reflect how we as human beings use the range of television programmes that are available to us, largely irrespective of just how these programmes physically reach us. This constancy of how we view is the justification for the limited update of this book. At the very least, the patterns described here will provide the base-line or norm.

1.3 The programmes available

On the traditional UK and US channels a variety of programmes tend to be screened. The selection and mixture will vary with the aims and financial viability of the programme company or station. Where advertising provides the revenue, a broad tendency is to aim at maximum audiences, subject in the UK to some public accountability or control for balance by the IBA. But even a broadcaster like the BBC which is independent of advertising revenue is conscious that one measure of its success is the size of the audiences it can attract and hold.

A broad comparison by type of programme brings out certain differences in the 'programme mix' of different channels, as in Table 1.1 (adapted from Williams, 1974).

Repetitive programming

Perhaps the dominant feature of television programming is that it is repetitive. The same programmes – or strictly speaking, different episodes of the same programme – tend to be shown at the same time each week, with perhaps two or three major upheavals in the programme schedule a year.

Some of this repetitive programming takes the form of serials with on-going story-lines and a strong element of continuity. Others are series, film or drama slots, regular current affairs programmes, week-end sports broadcasts and of course the news, where the need to watch regularly would appear to be less.

Repetitive programming superimposes a firm structure on television. The number of occasions where there is something radically new to watch are relatively few. There are several reasons for this.

Firstly, television demands an enormous amount of programme material. For everything to be original and new would require the development and organisation of experienced talent that is hardly likely to be available. Proverbially the film 'Lassie' was followed by 'Son of Lassie'.

Secondly, we like repetition. As viewers we find it convenient to know in advance what programmes are being shown and when – even if we are only going to miss them. We also learn to appreciate certain characters, comedy

Table 1.1
The distribution of programmes on various channels
(by time of duration)

March 3–9, 1973	Commercial		Public Service		
	ITV UK*	Ch7 US**	BBC1 UK	BBC2 UK	KQED US**
Programme Type	%	%	%	%	%
Series and Serials	17	17	7	4	5
News & Public Affairs	13	14	25	12	22
Movies	12	18	6	11	6
Education	12	2	23	29	26
Commercials	11	14	–	–	–
Gen. Entertainment	10	24	7	7	0
Childrens' programmes	8	4	11	6	27
Features and Document.	6	1	7	20	6
Sport	6	4	6	2	2
Plays	3	0	5	5	0
Publicity (internal)	1	1	1	1	1
Religion	1	1	1	0	0
Arts and Music	0	0	1	3	5
Total hours transmitted	103	133	100	62	94

Note: *Anglia **San Francisco

routines or stereotype plot situations through familiarity. We develop habits and preferences.

Thirdly, on commercial television fairly stable and predictable audience levels are required at each point of time so that advertisers have some idea of what future audience sizes they are buying.

Repetitive programming also occurs at closer than weekly intervals. The outstanding example is the daily news. But some series in the UK are shown twice weekly or more often, especially some on week-day afternoons for young children in specific age-groups. In the US the smaller 'independent' stations, and also sometimes the networks during day-time, go in for 'strip-programming'. This means that episodes of the same programme (often old ones) are shown regularly on each of the week-days (e.g. 'The Lucy Show', 'The Flintstones' and 'Star-Trek' five times a week each).

Despite this largely repetitive nature of TV programming, little had been published about the relatively low extent (generally some 50% or less) to which viewers of one episode of a programme also view the next. This is discussed particularly in Chapter 5.

1.4 Audience measurement

The analyses of the television audience discussed in this book are generally
based on routine audience measurement data. Broadcasting is unique in the
extent to which it has to rely on market research to indicate its uptake. Other
media and industries may use circulation figures, box-office receipts or
ex-factory shipments to estimate sales or even the number of customers. But
for television the only currently practical way to determine which programmes
are viewed or favourably evaluated is to conduct sample surveys or the like.
This produces the familiar audience size-ratings, i.e. measures of the
percentage of the total population viewing.

Audience size

In the UK the main audience measurement system is sponsored by BARB (the
Broadcaster Audience Research Board) which is made up of ITV programme
companies and the BBC. The work is currently carried by AGB (Audits of
Great Britain).

The data are based on panels of households whose viewing is measured more
or less continuously over successive days and weeks. Sample sizes tend to be
from 100 up to 500 homes per region, totalling about 3000 nationally and
comprising some 7000 individuals aged 4+. Individuals' viewing has tradition-
ally been measured in quarter-hour time-bands by weekly diaries, the
qualifying definition being that one has viewed for at least 8 minutes in the
quarter-hour. The diary data is monitored by an electronic meter attached to
the TV set, which itself gives minute by minute 'set-on' ratings.

These procedures tend to give reliable results. The quarter-hour measurements of individuals' viewing are the data mainly used in this book. Since late
1984 the viewing of individuals has been measured by push-button
'people-meters' rather than diaries and the 'viewing' criterion has been
reduced to 1 minute. Rather higher audience sizes are being recorded. (For a
somewhat fuller history of audience measurement until 1975, see Appendix A
of the first edition.)

In the US the audiences of the national network programmes are mainly
measured by a national meter panel operated by the A.C. Nielsen Company
and supplemented by separate diary measurements of individual viewers'
behaviour. Audience ratings for local 'markets' are measured by both the
American Research Bureau (Arbitron) and Nielsen, using one-week diaries.
There are also now local 'overnight' meter panels in some of the major
markets.

Audience appreciation

Audience size by itself is an incomplete index of viewers' reaction to the programmes on offer. In Chapters 8 and 9 we discuss results from more attitudinal measurements, e.g. from 'Audience Appreciation' panels operated in the UK by the IBA and now also by BARB. Samples of viewers report with mail diaries the programmes they have viewed in terms of an overall Appreciation Index (running from 'Not at all interesting and/or enjoyable' to 'Extremely interesting and/or enjoyable').

1.5 Summary

In developed countries like the UK and the US most people can choose from different television programmes shown on two or more channels, each tending to carry a variety of different types of programme.

The dominant feature of television programming is probably its repetitive nature, with different episodes of the same programme often being shown at the same time in successive weeks.

2 Audience Flow

In this chapter we introduce two main aspects of audience flow. One is the extent to which viewers of one programme are also viewers of another programme ('audience duplication'). The second is the extent to which viewers also watch another episode of the same programme, usually shown a week later ('repeat-viewing').

2.1 Characterising the audience

We characterise the viewers of a particular TV programme here by what other TV programmes they also watch. The reasons for this approach are several.

First, it seems relevant to ask whether the people who watch a certain western also watch many other westerns but not current affairs programmes say. That tells us about the kind of viewers they are. Secondly, much data for such analyses exists so that any findings can be well substantiated. Thirdly, simple and generalisable results have in fact been established. Finally, the same approach is applicable in the future, whatever the technological changes say, and in different countries.

Other possible ways of describing the audience of a particular programme are by the viewer's age, sex, social class, occupation, size of household, number and age of children, ownership of a motor car or cheque book and so on. Viewers can also be differentiated by their attitudes, by measures of their personalities, their tendencies towards violence or sloth (both possibly induced or aggravated by television) and by their other activities and interests.

These other approaches have been followed in a number of studies of television audiences. For example, in the UK women tend to watch perhaps 10% more TV on average than do men, and older people much more (50%) than younger ones. But otherwise these approaches have produced few very revealing or insightful results (e.g. Frank and Greenberg 1980) and thus will not be pursued extensively here. Television viewing is largely a mass market activity.

Viewers can also be questioned about their appreciation or liking of

8

particular television programmes. But such attitudinal responses become much more telling if their interpretation is linked to what people actually do by way of choosing programmes to watch. This will be considered in Chapters 8 and 9.

2.2 Interpretative norms: high, low or normal?

Viewing patterns may be thought difficult to interpret because so many different factors can influence them. If only 52% of the viewers of a certain western also watch the Friday 'News at Ten', this could be because some of them were not at home then or had gone to bed, because of alternative programmes on other channels, because of what other members of the family or guests were doing, because there happened to be no particularly newsworthy news that day, or even because some do not like watching the news. So how do we judge this figure?

If we were told that 52% of the viewers of the western read *The Times*, we might find that easier to interpret. We would already know that such a figure is very high compared with the percentage of the population as a whole who read *The Times*, and also that readers of *The Times* tend to be 'serious' readers. In contrast, it generally would not be immediately obvious whether a figure of 52% of the western audience watching the news should be considered high, or low, or perhaps just normal.

Thus what is needed in the first place are some interpretative 'norms', i.e. typical patterns of audience duplication and repeat-viewing for different programmes. Once we determine these, individual figures like the 52% should be easier to interpret. It is our concern in the next few chapters to describe and discuss such general patterns.

2.3 Audience duplication

One basic aspect of viewing behaviour is the extent to which viewers of one programme also watch another programme. This may be screened on a different day and perhaps on a different channel. There are many possible factors involved – time of day, day of week, channel, type of programme, programme content, audience size or 'rating' of each programme and so on. What are the patterns and the factors that influence the duplication of audiences for different programmes? Sufficient research has taken place to provide relevant results. One basic result is known as the duplication of viewing law.

The duplication of viewing law

The major influence on the level of audience duplication between two programmes is usually the rating level or audience size for each programme. This is to some extent self-evident. If almost nobody watches a programme, audience duplication with any other programme will be low! But the facts show that the effect of rating levels is pervasive even in less extreme cases. It can be summarised by the duplication of viewing law. This states that:

> The proportion of the audience of any TV programme who watch another programme on another day of the same week is directly proportional to the rating of the latter programme (i.e. equal to the rating, times a certain constant).

Thus, given knowledge only of the ratings of two programmes on different days and the value of the constant, we can predict what proportion of the audience of the first programme will also watch the second; the audience overlap simply varies with the rating level (the audience size) of the latter.

For example, if the constant or 'proportionality factor' for ITV programmes is 1.7 and the rating of the second programme was 20 (i.e. 20% of the total population watched it), then the duplication law states that about

$$20 \times 1.7 = 34\%$$

of the audience of the *first* programme will have watched the second.

This numerical example brings out the dramatic feature of the result. It implies that the proportion of the audience of any ITV programme on one day who saw the ITV programme on the other day is about 34%, i.e. that the audience overlap generally depends only on the proportionality coefficient (here 1.7) and the latter programme's rating (here 20) and *not* on the content of either programme. This will be shown in detail in the following chapters.

The result might seem surprising, e.g. that the programme would be seen by about 34% even of the audience of a children's programme shown on a previous afternoon. But if we are talking about adults, few would have watched the children's programme and about 34% of these might well then watch the other programme. That is no longer so surprising.

The duplication law can also be expressed more symmetrically, i.e. not as the proportion of the audience of programme A say who also watch programme B, but as the percentage of the whole population who watch both A and B. The law then takes the form:

% of population who watch both A and B
= rating of A × rating of B × the proportionality coefficient divided by 100

(see also the appendix to this chapter for the algebra.)

Thus if programme A had a rating of 30 and programme B one of 20 as before, then with a proportionality factor of 1.7, about:

$$\frac{30 \times 20 \times 1.7}{100} = 10.2\%$$

of the population will have watched both programmes. (The divisor of 100 is needed if we are using percentages and not proportions. We note that 10.2 out of 30 is 34%, as before.)

Empirical basis

The duplication of viewing law may be difficult to accept at first sight. But it has held under many different circumstances, i.e. it is an empirical generalisation. It describes, to a close degree of approximation, what actually happens. It also has a theoretical basis or explanation which will be discussed in Chapter 5. One of the tasks of this book is to illustrate how and when the law works and also to pinpoint and discuss those specific situations where audience flow takes other forms.

The law generally accounts for the single major factor in audience duplication – audience ratings. The influence of other factors then shows itself as deviation from the predicted results. These deviations tend to be small. But they could not be isolated without first accounting for the effects of the rating levels as such.

There are certain specific situations where the law does not directly apply. Thus, the audience overlap for late-night programmes is consistently higher than the duplication law predicts because the same people tend to stay up late. Again, two programmes shown in succession on the same channel have a much higher degree of audience duplication because of the 'inheritance effect', as discussed in Chapter 4. (Hence the earlier emphasis on programmes shown on *different* days.)

Acceptance of the duplication of viewing law as a useful summary depends not on any purely theoretical argument or assumption, but only on its ability to fit the facts. This is more fully discussed in Chapters 3 and 4.

An illustration: 'arts documentaries'

We now use an early example to illustrate how the duplication of viewing law can be used to test whether certain types of programme attract particular types of habitual viewer. If such a tendency were to exist for that programme type, then the proportions of the population watching pairs of such programmes should be higher than the predicted value given by the duplication law.

For example, do 'arts documentaries' (using the formal IBA definition of programme-type categories) attract a steady following of 'cultured' viewers? In the first week of May 1967 five such programmes were offered on the ITV and BBC1 channels, as shown in Table 2.1.

Table 2.1
Viewing of 'arts documentaries'
(adults, May 1967)

Channel	Time (pm) & Day	Programme	Rating
ITV	9.15 Wednesday	Cinema	43
ITV	10.15 Friday	This Week – the Arts	12
BBC1	11.00 Wednesday	Masterwork – Piano	1
BBC1	10.15 Sunday	Contours of Genius	5
BBC1	11.00 Sunday	Look of the Week	2

The duplication law predicts that the percentage of the total adult population who watched both 'Cinema' and 'This Week – the Arts' is the product of the audience ratings multiplied by the *within-channel* constant, at that time 1.4, divided by 100, i.e.:

$$\frac{43 \times 12 \times 1.4}{100} = 7.2\%$$

(The value of the duplication-constant 20 years ago, at 1.4, was somewhat lower than now, as described in Chapters 3 and 4.)

Examination of actual viewing data showed that 8.0% of the population had in fact watched the two programmes. This is close to the prediction of 7.2, with a discrepancy of only 0.8, or 1 percentage point to the nearest whole number.

It may be thought that 'Cinema' and 'This Week – the Arts' are rather different in terms of content. But similar comparisons can be made for all the other pairs of programmes in the broad 'arts' category. For example, the duplication for ITV's 'Cinema' and the BBC's 'Contours of Genius' was about 1 percentage point *below* the estimated BBC × ITV norm. That for ITV's 'This Week – the Arts' and 'Contours of Genius' was about 1 rating point *above* the predicted level. And so on.

Taking all the eight pairs of programmes that were shown on different days, the discrepancies were all 1 rating point or less with no systematic tendency up or down. The sequence of discrepancies (observed minus predicted duplication) to the nearest whole number was

$$0, -1, -1, 1, 1, 0, 0, 0$$

Thus the average discrepancy was zero and there certainly was no general tendency for viewing of different pairs of arts programmes to be substantially higher than predicted by the general law. Instead, the duplications were close to the predicted levels. People who watched one arts programme had no more tendency to see some other arts programme than to see, say, a western or a religious programme or a sports programme with a comparable rating. The pattern of viewing was therefore not related to programme content.

This result illustrates a practical application of the kind of findings reported here. Analyses of past viewing patterns over thousands of programmes have provided us with a result (here the duplication law) which enables us to predict successfully the proportion of people who will see any two programmes. Hence we can test whether programmes of a certain type attract an exceptionally large number of viewers in common. With the 'arts documentaries' typically there was no such tendency. The observed duplications were virtually the same as those for any kinds of programme.

2.4 Repeat-viewing

A different pattern occurs for repeat-viewing, i.e. the extent to which the same people view different episodes of the same programme, usually shown weekly, but sometimes (as in the USA in the afternoons) daily.

In most cases the size of the audience for successive episodes tends to be much the same. Table 2.2 illustrates this for four US programmes which were shown on the ABC network over all five week-day afternoons in New York City in a typical week during 1974.

Table 2.2
Ratings for repetitive programmes on five week-days

New York Housewives Jan–Feb 74		% HW's Viewing					
		Mon	Tue	Wed	Thu	Fri	**Av.**
ABC							
4.00	Love Amer. Style	3	2	4	3	3	**3**
4.30	4.30 Movie	4	4	4	4	4	**4**
6.00	Eyewitness News 6	11	10	13	12	11	**11**
7.00	ABC Evening News	8	8	10	8	8	**8**
Average		7	7	8	7	7	**7**

The ratings are clearly much the same on the different days, with Wednesday just fractionally higher. About 3 or 4% of all housewives watched 'Love

American Style' each day, about 4% 'The 4.30 Movie', about 11% the 6 pm
'Eyewitness' news and 8% the main 'ABC News'.

This steadiness of the ratings day by day might suggest that it is very much the
same people watching at a given time each day. But this is largely not so – on
average only about *half* the audience on one day also watches the same
programme on another.

Table 2.3 illustrates this in terms of the percentage of the Monday–Thursday audiences for each programme who also watched the Friday episode. Most
of the repeat-viewing percentages are only in the 50s or 40s, with that for 'The
4.30 Movie' slot even lower.

Table 2.3
Repeat-viewing within the week
(% of Monday to Thursday audiences also watching on
Friday)

New York Housewives Jan–Feb 74	% Viewing on Friday of the Audience on				
	Mon	Tue	Wed	Thu	**Av.**
ABC					
4.00 Love Amer. Style	38	47	39	56	**45**
4.30 4.30 Movie	29	26	43	51	**37**
6.00 Eyewitness News 6	59	55	56	58	**57**
7.00 ABC Evening News	52	50	43	49	**48**
Average	45	44	45	57	**47**

Most television programming is repetitive week by week and here audience
sizes for successive episodes tend also to be fairly steady. For example, roughly
15% might watch 'Panorama' and about 25% 'Star Trek' each week. Despite
this relative steadiness of the ratings, the facts show that the general level of
repeat-viewing in the UK is also only about 55% or so.

One may here be dealing with a *serial* (i.e. a programme with a continuing
story-line) or a *series* (i.e. the same characters appearing in episodes with
self-contained plots, or comedy or variety shows with the same star
performers). Less continuity occurs with drama slots ('The Monday Play'),
screenings of old films ('The Tuesday Film'), regular current affairs programmes ('Midweek') or the news. These various programme factors might be
expected to influence the extent to which people view a programme regularly.
But the findings show that these factors appear to matter little. In general,
repeat-viewing levels for different types of programme all tend to be about
55%. More detailed results are discussed in Chapter 5.

2.5 Rating levels

The outcome of any study of viewing patterns and audience flow might be expected to improve our understanding of why some programmes achieve higher ratings than others. But at present this is not so. If anything, the reverse is true. Thus in examining audience duplication for different programmes, the primary factors are usually the rating levels themselves: given the ratings of two programmes we can successfully predict what proportion of the audience is in common.

In general not much is known by way of systematic and generalisable results about the factors which make one programme of a certain type more popular than another of the same type. There are apparently no well-established empirical findings or theoretical models from which ratings could be successfully predicted, nor can they even be systematically unravelled with hindsight. Predicting audience appeal and rating level is an area which is still very largely a matter of practical experience and judgement, in as far as it can be done at all.

2.6 Some statistical considerations

The results discussed in this book are all derived from sample data. Many of the illustrations are based on relatively small panels of, for example, 350 or so housewives in London. With samples of this kind, statistical sampling errors will occur in any particular case. These show themselves in terms of irregularities in the results.

However, the main findings reported here are essentially *regularities*. These could not arise as sampling errors since our regularities have generally been found to hold for a great variety of different samples, e.g. housewives and men, people in different regions and data for different years. We therefore do not have to worry about the statistical 'significance' of the main results, such as the duplication of viewing law itself or the relatively steady '55%' repeat-viewing levels that are reported.

Problems of statistical significance arise only with *discrepancies* from these regularities (e.g. Ehrenberg 1982, Chapter 10). Then the question is whether any particular discrepancy is mainly due to a 'sampling error' (i.e. it would not have occurred in other similar samples), or whether it would also have shown up in other different samples or in a much larger sample and hence be something to consider further.

The crucial question is whether such deviations can then themselves be seen to form some generalisable sub-pattern. When they do (e.g. the deviations from the duplication of viewing law for late-evening programmes mentioned earlier), they are highlighted in the discussion.

Many of the illustrations in this book are based on a sample of housewives in London in 1974 (housewives are convenient as there is only one principal one per household). The resulting data then refer to all housewives in households in the London ITV area with sets capable of receiving the relevant transmissions (ITV and BBC1 at the time). That is the 'population' in question in such a case. (Guest viewing is generally ignored in this book as the data cannot show whether guests at different times are the same people.)

But many other population sub-groups have also been covered in the analyses, as well as data for other UK regions and for the US, and for the late seventies and early eighties. The main justification of the results is their generalisability to different places and to different points in time and, hence, to different programme availability.

The definition of a rating

We often talk about the 'rating of a programme' or the 'viewers of a programme', but the data strictly refer to a quarter-hour time-band (watching for at least 8 minutes qualifies as a 'viewer') and not to the whole programme. People usually watch a whole programme. In the case of half-hour programmes, about 95% of those who watch the first quarter-hour also watch the second. With longer programmes more erosion of the audience occurs – up to about 20% of initial viewers may be lost by the end and even more late in the evening. More attention will have to be paid to the precise definition of the viewing audience if one is literally concerned with the programme as an entity.

2.7 Summary

In this chapter we have introduced the notion of audience flow – the extent to which the same people watch different television broadcasts. There are two main results. One establishes that generally only about 55% of the viewers of one episode of a programme watch the next episode of that programme. The second, the 'Duplication of Viewing Law', states that the size of the audience common to two different programmes on different days depends on the ratings of the programmes and the channels on which they are shown and not on the content of the programmes.

We proceed to a more detailed discussion of these findings in the following chapters.

Appendix: the duplication law

For the mathematically inclined we state the duplication of viewing law symbolically. For two times (or programmes) s and t, the law in its symmetrical form reads

$$r_{st} = kr_sr_t$$

Here r_{st} is the proportion of the audience watching both at times s and t, r_s and r_t are the proportions watching at each of these times, and k is a coefficient whose value is generally the same for different pairs of programmes.

Separate values of the coefficient k can be calculated for each pair of programmes, by the ratio r_{st}/r_sr_t of the observed values of r_{st}, r_s and r_t. If the resulting values of k for different cases are all more or less the same, i.e. about constant, then the law with a *constant* value of k does in fact hold across this range of cases, as a close approximation. This single value of k can then be estimated as the average of all the individual values of k.

A statistically more robust estimate of k is usually derived by forming the ratio of the total of the duplicated audiences for all pairs of times in question (i.e. Sum r_{st}) to the total of the cross-products r_sr_t for all pairs of times (i.e. Sum r_sr_t), namely

$$k = \text{Sum}(r_{st})/\text{Sum}(r_sr_t)$$

The above version of the law is useful when analysing large amounts of data. The alternative formulation of the law mentioned earlier in this chapter consists of dividing the equation $r_{st} = kr_sr_t$ by r_t, referring to the proportion of those viewing at time t who also view at time s. This reads

$$\frac{r_{st}}{r_t} = kr_s$$

This version is more helpful in allowing one to see, or demonstrate, the numerical pattern in the data. Thus r_{st}/r_t should depend only on the rating at time s, i.e. be the same for different times t. This constancy is then easier to look for in empirical data.

If all the ratings are expressed as percentages rather than as proportions, the symmetrical form of the duplication law has to be written as $r_{st} = kr_sr_t/100$ since each r value is multiplied by 100. But in percentage terms we still have $r_{st}/r_t = kr_s$.

Example

If the observed value of r_{st}, the audience common to s and t, is 6.5% and the ratings at s and t are 25 and 15, then the estimated value of k is $(6.5 \times 100)/(25 \times 15) = 1.73$, or 1.7 for short. If for other pairs of times the value of k is also about 1.7, the duplication law is operating with $k = 1.7$.

Conversely if we already know that k is generally about 1.7 and we have ratings $r_s = 25$ and $r_t = 15$, then the *predicted* duplicated audience is

$$1.7 \times 25 \times 15 \div 100 = 6.4$$

In other words 6.4% of the population should watch *both* programmes.

As a proportion of the audience at time t, the observed duplication $6.5/15 = 0.43$ or 43%, which equals 25×1.7. Thus 43% of the audience at time t watch at time s. This is somewhat less than twice as high ($k = 1.7$) as the proportion of the total population who watch at $s(r_s = 25\%)$.

Similarly, the proportion of the audience of s who watch at time t is $6.5/25 = 26\%$, which is about 1.7×15. The number of viewers at time t who also watch at time s is the same as the number of viewers at s who also watch at t (indeed, they are the same people). But this duplicated audience is a greater proportion of the smaller audience at time t (43%) than of the larger audience at time s (26%).

3 Viewing Different Programmes

In this chapter we examine the extent to which viewers of one programme also watch another – the overlap or 'duplication' between the audiences of different pairs of programmes.

The basic question is, 'How likely are viewers of one programme to watch another programme?' A pattern emerges within the context of the duplication of viewing law. The argument involved is not complex but it is rather detailed. We therefore first summarise the main steps, using some results from 1971 as illustrations.

3.1 The effect of the channels

People who saw a particular ITV programme yesterday are more likely than those who did not to watch a given ITV programme today. This tendency does not depend on the specific programmes or days in question but holds more generally; it also occurs for programmes on non-consecutive days and for ones in different weeks. Much the same pattern occurs on BBC channels, where viewers of one BBC programme are more likely than its non-viewers to watch a BBC programme on another day.

There is, however, no such positive tendency for audience flow *across* channels. Viewers of an ITV programme are if anything slightly *less* likely than its non-viewers to watch a given BBC programme. Similarly, viewers of a BBC programme are slightly less likely than its non-viewers to watch a programme on ITV.

These findings therefore point to a certain degree of channel loyalty, although it is not very large. Viewers on one channel are *somewhat* more likely to view that channel again than a specific other channel. (But they are also quite likely to watch a programme on that other channel.) The strength of the preference varies little for the different channels in Britain. A similar pattern occurs in the US for the three national networks (as is discussed in Chapter 7), but with slightly less loyalty to each particular channel than in the UK.

19

The detailed results hang together quantitatively in terms of the duplication of viewing law introduced in the last chapter. The proportionality coefficient of this law reflects the extent of channel loyalty. Some numerical examples now serve to illustrate this.

3.2 Two ITV programmes

We start with two specific ITV programme transmissions in London in April 1971. They were: 'The Mind of J.G. Reeder' (9 pm Monday 19 April) and 'The Adventurers' (8 pm Tuesday 20 April). These two programmes had similar ratings amongst London housewives: 30% of the housewives saw 'The Mind of J.G. Reeder' and 28% saw 'The Adventurers'. (These were quite high ratings as there were only two channels then.)

The proportion of the population the two programmes had in common is established by cross-tabulating the viewing data. This shows that 14% of London housewives saw *both* of the programmes:

% of housewives seeing 'J.G. Reeder'	30
'The Adventurers'	28
both	14

This means that about half of the audience of one programme (14 out of 30 or out of 28) also saw the other. This level of audience overlap is high compared with the ratings of either programme. Thus Table 3.1 illustrates that 'The Adventurers' was seen by 46% of the viewers of 'J.G. Reeder', compared with only 20% of the *non-viewers* of 'J.G. Reeder' and 28% of all London housewives. (The latter figure – the rating – lies closer to the percentage among the non-viewers of 'J.G. Reeder' because there are more non-viewers of 'J.G. Reeder', 72%, than viewers 28%.)

We can therefore conclude that there was a special tendency for viewers of

Table 3.1
The audience overlap between the two programmes

London Housewives April 1971	% who viewed The Adventurers on Tuesday
Viewers of J.G. Reeder on Monday	46
Non-Viewers of J.G. Reeder	20
All Housewives (the rating)	28

the Monday ITV programme to view the one on Tuesday as well. Table 3.1 shows that they were more than twice as likely as the non-viewers of the Monday programme to do so.

3.3 Other ITV programmes

The question now is whether this high duplication of the two audiences reflects a general pattern or only something peculiar to the two programmes, or the times at which they were shown, or to some other specific factors. Both 'J.G. Reeder' and 'The Adventurers' were episodes in fictional series. The high audience duplication between them could therefore simply have been due to a special inclination amongst certain people to watch programmes of this particular type.

However, Table 3.2 shows that 'The Adventurers' on Tuesday was also watched by between 40 and 50% of the housewife audiences of other 'peak-time' ITV programmes on Monday evening. Thus 'The Adventurers' was just about as popular with Monday viewers of 'World in Action' (48%) at 8 pm and of 'News at Ten' (43%) at 10 pm as with viewers of lighter programmes such as 'Opportunity Knocks' (50%) at 7 pm and 'J.G. Reeder' itself (46%) at 9 pm.

Table 3.2
The audience overlap between various Monday programmes
on ITV and 'The Adventurers'

	% who viewed
London Housewives April 1971	The Adventurers Tuesday 8 pm
Viewers of	
Opportunity Knocks – Monday 7 pm	50
World in Action – Monday 8 pm	48
J.G. Reeder – Monday 9 pm	46
News at Ten – Monday 10 pm	43
All Housewives (the rating)	28

It therefore appears that the high audience duplication observed between 'J.G. Reeder' and 'The Adventurers' was not a function of any similarity between these two programmes as such, nor of the particular times at which they were shown. Instead, it seems to reflect a *general* tendency for the viewers of any ITV peak-time programme on Monday to have watched 'The

Adventurers' on Tuesday: 40 to 50% did so, compared with only 28% of the housewife population as a whole.

We now have to check whether this high audience overlap was specific to 'The Adventurers' or also applied to other peak-time ITV programmes on Tuesday. Table 3.3 shows that other ITV peak-time programmes on Tuesday were also watched by at least 40% of the audiences of the various Monday programmes. Thus 'Bless This House' at 7 pm on Tuesday was watched by more than 50% of any of the four previous evening's audiences, 'The Saint' at 9 pm on Tuesday by between 40 and 50% and so on.

Table 3.3
High overlap between Monday and Tuesday ITV programmes

London Housewives April 1971	% of Monday Audience who on TUESDAY Viewed			
	Bless This House 7 pm	The Adventurers 8 pm	The Saint 9 pm	News at Ten 10 pm
MONDAY				
Opportunity Knocks – 7 pm	69	50	41	45
World in Action – 8 pm	60	48	43	44
J.G. Reeder – 9 pm	53	46	46	46
News at Ten – 10 pm	53	43	39	53
Average	59	47	42	47
All Housewives (the rating)	33	28	23	29

All these overlap figures are high compared with the ratings of 23–33% of the Tuesday programmes themselves, shown at the bottom of Table 3.3. Thus markedly higher proportions of the viewers of any of the Monday ITV programmes watched each of the Tuesday ITV programmes.

3.4 The duplication of viewing law

The column averages in Table 3.3 vary from a high of 59 for 'Bless This House' (7 pm Tuesday) to a low of 42 for 'The Saint' (9 pm Tuesday). This variation in the overlap audiences largely mirrors the variation in the ratings themselves which range from 33 for 'Bless This House' down to 23 for 'The Saint'. The audience overlap or duplication tends to vary proportionally with the rating of the Tuesday programme. Table 3.4 shows this relationship more graphically by arranging the four Tuesday programmes in descending order of their ratings.

Table 3.4
Tuesday programmes in order of their ratings

London Housewives *April 1971*	% of Monday Audience who on TUESDAY Viewed				
	Bless This House	News at Ten	The Adventurers	The Saint	Average
Av. Duplication (T3.3)	59	47	49	42	49
Housewife Rating	33	29	28	23	28
Av. Dup./Rating	**1.8**	**1.6**	**1.7**	**1.8**	**1.7**

The ratios of the average duplication to the rating are very similar, ranging only from 1.6 to 1.8 and averaging about 1.7. We therefore can say that to a quite close degree of approximation the percentage of the viewers of a Monday ITV peak-time programme who watched a Tuesday ITV programme is about 1.7 times the percentage of the total housewife population who watched that programme. That is, the duplicated audience is generally higher than the rating by about 70% of the latter.

Table 3.5 compares these theoretical estimates (1.7 × ratings) with the average observed duplication levels for each programme.

Table 3.5
The theoretical estimates 1.7 × rating

London Housewives *April 1971*	% of Monday Audience who on TUESDAY Viewed				
	Bless This House	News at Ten	The Adventurers	The Saint	Average
Average Duplication	59	47	49	42	49
1.7 × Rating	**56**	**49**	**48**	**39**	**48**

The agreement is clearly close, generally within a few percentage points. This is a case of the duplication of viewing law described in the last chapter. It has been found to occur in many thousands of cases, as will be illustrated further on in this book.

The theoretical estimates of 1.7 × rating do not, however, agree quite as closely with the observed audience duplication between the individual Monday and Tuesday programmes that were shown in Table 3.3. There are additional factors involved in these duplication patterns (including sampling errors, since the data are based on samples in the hundreds rather than the thousands). The largest discrepancy is for the overlap between the two 7 pm programmes, where the observed duplication is 13 percentage points higher than predicted. This is a systematic feature to which we shall return. Otherwise the discrepancies are on average only about 3 percentage points. Thus we can say that the main pattern in the data shows that the duplication levels are about 70% higher than the ratings.

3.5 Viewing on non-consecutive days

This high level of audience overlap between Monday and Tuesday programmes might occur because we have analysed viewing on consecutive days. It could be that TV sets tend simply to be kept tuned to the channel last watched the night before. However, further analyses show that the pattern illustrated so far, and the duplication of viewing law in particular, also holds for audiences on *non-consecutive* days.

Table 3.6 illustrates this for Monday programmes and ones on Wednesday, two days later. Once more the percentage of Monday ITV viewers watching an ITV programme on Wednesday is about 70% higher than the latter's rating (i.e. the average duplication equals 1.7 times the Wednesday rating). For example, 35% of all housewives watched 'This is Your Life' at 7 pm on Wednesday (the rating), but about 60% of the Monday audiences did so (with again an abnormally high value of 69% for the two 7 pm programmes, just as in Table 3.3).

Table 3.6
Duplication of viewing between Monday and Wednesday ITV programmes

| London Housewives April 1971 | % of Monday Audience who on WEDNESDAY Viewed | | | |
	This is Your Life 7 pm	I Spy 8 pm	Hine 9 pm	News at Ten 10 pm
MONDAY				
Opportunity Knocks – 7 pm	69	60	53	50
World in Action – 8 pm	62	62	50	49
J.G. Reeder – 9 pm	55	57	50	49
News at Ten – 10 pm	55	54	51	55
Average	60	58	51	51
1.7 × Rating	**60**	**61**	**54**	**53**
Rating	35	36	32	31

Table 3.7 illustrates a further result for the Monday programmes and ones on a Tuesday *two weeks later*. The pattern is much the same as before, but using a coefficient of 1.7 in the duplication law tends to overstate the observed

results by a few per cent. A coefficient of 1.6 would give a closer fit. It is not yet clear whether this suggestion of a slight erosion of duplication levels is general.

Table 3.7
Duplication for the Monday ITV programmes and those on Tuesday TWO WEEKS LATER

London Housewives April 1971	*% of Monday Audience who Viewed on TUESDAY (3 May)*			
	Bless This House 7 pm	The Adventurers 8 pm	The Saint 9 pm	News at Ten 10 pm
MONDAY (19 APRIL)				
Opportunity Knocks – 7 pm	51	55	42	40
World in Action – 8 pm	45	50	43	41
J.G. Reeder – 9 pm	46	50	46	42
News at Ten – 10 pm	42	45	38	40
Average	46	50	42	41
1.7 × Rating	**44**	**54**	**48**	**48**
Rating	26	32	28	28

3.6 Channel switching

We have now seen that there is a relatively high overlap between the audiences of peak-time ITV programmes on different days. Viewers of one ITV programme are substantially more likely than non-viewers of that programme to watch another ITV programme.

This high duplication of viewing for ITV programmes need not by itself imply any form of 'loyalty' to the ITV channel. For example, viewers of an ITV programme could also be more likely than non-viewers to watch a programme on another channel. They may simply be heavy viewers of television.

The high duplication for the ITV programmes implies special 'loyalty' to the ITV channel only if such high audience duplication does not occur *between* channels, e.g. between the audience of an ITV programme on the one hand and that of a BBC1 programme on the other.

In practice, between-channel audience overlap is in fact relatively low. This is illustrated in Table 3.8 for the same Monday and Tuesday that we have already been examining (19 and 20 April 1971). The data show the extent to which viewers of the ITV peak-time programmes on the Monday watched the BBC1 peak-time programmes on the next day.

Table 3.8
ITV versus BBC1: the tendency for Monday's ITV audiences to view
BBC programmes on Tuesday

| | % of Monday ITV Audience who on TUESDAY viewed BBC1's | | | |
London Housewives April 1971	Top of the Form 7 pm	A Shot in the Dark 8 pm	News 9 pm	Civilisation 10 pm
MONDAY ITV				
Opportunity Knocks – 7 pm	12	23	24	7
World in Action – 8 pm	12	24	26	6
J.G. Reeder – 9 pm	12	23	28	9
News at Ten – 10 pm	9	25	29	9
Average	11	24	27	8
0.8 × Rating	**15**	**22**	**25**	**9**
Rating	19	28	31	11

Thus 'Top of the Form' at 7 pm on Tuesday on BBC1 was watched by between 9 and 12% of the viewers of any one of the Monday ITV programmes; 'A Shot in the Dark' at 8 pm on BBC1 was watched by about 24% of the Monday ITV viewers and so on. These duplication levels are lower than the ratings of the BBC programmes themselves. As shown in the bottom line of Table 3.8, about 19% of London housewives watched 'Top of the Form', at 7 pm whereas only 9 to 12% of the ITV viewers did so. Such differences, although generally not quite as big, also occur for the later BBC programmes. Fewer ITV viewers therefore watched the BBC1 programmes than the proportion watching in the population as a whole.

The pattern still follows the duplication of viewing law, within limits of a few percentage points. But the proportionality factor is very different. It was on average 0.8 for ITV and BBC1 in 1971 as compared with 1.7 for two ITV programmes (Table 3.4). A duplication coefficient of 0.8 means that viewers of the ITV programmes are 20% less likely to watch the BBC1 programmes than the population as a whole. Viewing of an ITV programme therefore appears to inhibit the viewing of BBC1 programmes on another day.

The same pattern holds the other way round. Viewers of BBC1 programmes are less likely than the general population to switch to ITV programmes on another day. This is illustrated in Table 3.9 for the same programme combinations as in Table 3.8.

We therefore have a general pattern that viewers of an ITV programme on one day, when faced with the choice between similarly rated ITV and BBC1 programmes on another day, are about twice as likely to watch an ITV

Table 3.9
BBC1 versus ITV: the tendency for BBC1 audiences on Tuesday to view
ITV programmes on Monday

| | *% of Tuesday BBC1 Audience who on MONDAY viewed ITV's* | | | |
London Housewives *April 1971*	Opportunity Knocks 7 pm	World in Action 8 pm	J.G. Reeder 9 pm	News at Ten 10 pm
TUESDAY BBC1				
Top of the Form – 7 pm	20	18	20	15
A Shot in the Dark – 8 pm	26	26	25	29
News – 9 pm	24	25	27	29
Civilisation – 10 pm	21	15	23	26
Average	23	21	24	25
0.8 × Rating	**26**	**24**	**24**	**26**
Rating	32	30	30	32

programme again as to tune to BBC1 that day (the duplication coefficients of 1.7 and 0.8 respectively). This is why the high duplication level between ITV programmes is in fact a reflection of channel loyalty.

3.7 Duplication on BBC1

So far our analysis of audience overlap for programmes shown on the same channel has been restricted to ITV, but the results for BBC1 programmes are similar. The general pattern is illustrated in Table 3.10.

There is some variability in the individual results, with an exceptionally high value occurring once more for the two 7 pm programmes. But it is clear that viewers of a Monday BBC1 programme are substantially more likely to watch a Tuesday BBC1 programme than is the population as a whole. The duplication levels are generally almost twice as high as the programme ratings, averaging at a factor of 1.8. This is typical of the more general run of results for BBC1.

The duplication of viewing law therefore also applies to BBC1 programmes and with a similar degree of overlap as for ITV – the coefficient in April 1971 was 1.8 for BBC1 programmes, compared with 1.7 for ITV programmes.

3.8 Duplication on BBC2

BBC2 programmes generally attracted smaller audiences than programmes on the other two channels. These lower rating levels might imply some special

Table 3.10
Audience overlap on BBC1

| London Housewives April 1971 | % of Monday BBC1 Audience who on TUESDAY Viewed BBC1's | | | |
	Top of the Form 7 pm	A Shot in the Dark 8 pm	News 9 pm	Civilisation 10 pm
MONDAY BBC1				
A Question of News – 7 pm	49	51	49	28
Panorama – 8 pm	46	48	56	23
News – 9 pm	39	45	55	28
Brett – 10 pm	31	43	47	21
Average	41	47	52	25
1.8 × Rating	**34**	**50**	**56**	**20**
Rating	19	28	31	11

pattern of viewing, e.g. viewers being more 'devoted' to the channel, or perhaps especially selective in what they viewed.

Viewing of BBC2 transmissions also follows the duplication of viewing law but in 1971 the apparent degree of audience overlap was much higher than on the other two channels. Table 3.11 illustrates overlap between the audiences of BBC2 programmes on different days (shown for the average pair of week-days because the sample sizes of homes able to receive BBC2 in April

Table 3.11
High duplication on BBC2
(The average for all pairs of week-days)

| London Housewives April 1971 | % of BBC2 Audience who on Another Day Viewed BBC2 at: | | |
	8 pm*	9 pm	10 pm
BBC2			
Average Week-day – 8 pm*	12	14	31
Average Week-day – 9 pm	11	17	25
Average Week-day – 10 pm	12	16	21
Average	12	16	26
3.2 × Rating	**13**	**16**	**26**
Rating	4	5	8

Note: *There were no 7 pm BBC2 transmissions

1971 were small). It shows that housewife viewers of a BBC2 programme on one day were about 3 times as likely to watch a BBC2 programme on another day as was the general population of housewives.

However, this high duplication is more apparent than real. There is a simple explanation. In 1971 a substantial number of housewives virtually never watched BBC2 because they did not have a set capable of receiving transmissions on this channel. Thus the duplication coefficient of 3.2 is high, not because channel loyalty is specially marked, but because the BBC2 ratings were depressed by this non-availability factor. Indeed, the duplication levels of about 10 to 30% shown in Table 3.11 are not high in absolute terms but only relative to the very low programme ratings.

The ratings of BBC2 programmes are therefore adjusted in Table 3.12 for 'availability', i.e. the percentage of housewives viewing the programme amongst those with sets able to receive BBC2 (about 60% of London homes in 1971). This reveals a pattern of audience overlap much more like that for ITV and BBC1. Compared with the higher ratings among homes with BBC2 sets, the duplication coefficient is now about 1.9, only slightly higher than the coefficients for the two main channels (1.7 and 1.8). The general pattern of channel loyalty in 1971 was thus broadly the same for all three channels. This was borne out subsequently when the ability to receive BBC2 increased to near 100%.

Table 3.12
BBC2 duplication compared with the ratings in BBC2 homes

London Housewives (with sets capable of receiving BBC2)	% of BBC2 Audience who on Another Day Viewed BBC2 at:		
	8 pm	9 pm	10 pm
Av. duplication (T.3.11)	12	16	26
1.9 × Rating	**13**	**15**	**25**
Rating (in BBC2 Homes)	7	8	13

3.9 Switching to BBC2

One of the arguments used in the sixties to support the award of the third TV channel to the BBC was that this would permit complementary rather than competitive programming. Instead of competing directly with ITV for audiences and hence providing similar 'popular' programmes, BBC2 was to offer a real alternative, providing viewers with a greater choice and satisfying more minority interests. With such a complementary programming strategy one might expect to find a special link between BBC1 and BBC2.

Table 3.13 illustrates the duplication pattern between BBC1 and BBC2 for the average pair of week-days in the week of 19 April 1971. It shows that the viewer of a BBC1 programme on one day tended to be fractionally more likely to watch BBC2 broadcasts on another day than was the average housewife with a BBC2-type set. The duplication coefficient is 1.1.

Table 3.13
Duplication between BBC1 and BBC2
(the average for all pairs of week-days)

London Housewives April 1971	*% of BBC1 Audience who on Another Day Viewed BBC2 at:*		
	8 pm	9 pm	10 pm
BBC1			
Average Week-day – 7 pm	11	11	12
Average Week-day – 8 pm	9	9	13
Average Week-day – 9 pm	9	8	12
Average Week-day – 10 pm	8	7	15
Average	9	9	13
1.1 × Rating	**8**	**9**	**14**
Rating in BBC2 Homes	7	8	13

The comparable figures for ITV viewers watching BBC2 on another day are set out in Table 3.14. Here the tendency is for an ITV viewer to be slightly *less* likely to watch a BBC2 programme than the average housewife in 1971. The duplication coefficient is 0.9.

The important aspect of these results is not the difference between the two duplication coefficients of 1.1 and 0.9 but their similarity. To a first approximation both coefficients are 1. Furthermore, they are not very different from the duplication level for BBC1 and ITV programmes seen earlier (a coefficient of 0.8).

Roughly speaking, viewers of either a typical BBC1 programme or a typical ITV programme were therefore just about as likely to watch a BBC2 programme on another day as was the average housewife with a BBC2 set. But BBC1 viewers tended to be about 10% *more* likely to do so and ITV viewers about 10% *less* likely. These differences therefore point to only a small degree of overall loyalty between the two BBC channels. But the link between BBC1 and BBC2 is far less marked than the loyalty towards either of the individual BBC channels or towards ITV.

Table 3.14
Duplication between ITV and BBC2
(the average for all pairs of week-days)

London Housewives April 1971	% of ITV Audience who on Another Day Viewed BBC2 at:		
	8 pm	9 pm	10 pm
ITV			
Average Week-day – 7 pm	6	7	12
Average Week-day – 8 pm	6	7	10
Average Week-day – 9 pm	6	8	14
Average Week-day – 10 pm	5	7	11
Average	6	7	12
0.9 × Rating	6	7	12
Rating in BBC2 Homes	7	8	13

3.10 The overall pattern

Table 3.15 gives the constants of the duplication of viewing law for April 1971, as they have been developed in this chapter and recurred among thousands of different programme combinations at that time.

Table 3.15
Constants of the duplication of viewing law in 1971

Within-channel duplication constants:	ITV × ITV	= 1.7
	BBC1 × BBC1	= 1.8
	BBC2 × BBC2	= 1.9
Between-channel duplication constants:	ITV × BBC1	= 0.8
	ITV × BBC2	= 0.9
	BBC1 × BBC2	= 1.1

The within-channel constants, all substantially above 1, show that channel loyalty existed for viewing on different days of the week (and across different weeks). BBC2 and BBC1 appear to have attracted a slightly more loyal following but the similarity of the various duplication constants is more remarkable than the marginal differences. The between-channel constants also show some slight BBC link, but again the similarities and closeness to a value of 1 are the striking points.

To summarise, in 1971 a peak-time programme with a rating of 20 was generally watched by just under 20% of the audience of any peak-time programme shown on a different channel on another day of the week, but by roughly 34% of the audience for another day's peak-time programme on the same channel.

3.11 The patterns over time

Two developments in the UK since 1971 that might have influenced channel loyalty were the spread of BBC2 – to the extent that almost all households can now watch the channel and most of them do at times – and the advent of Channel 4 in late 1982. But both changes have at most had very slight impacts on the duplication coefficients in the duplication of viewing law. (The figures for BBC2 came into line as the availability to receive it increased.) The basic pattern of audience duplication has remained the same.

Table 3.16 summarises this for the period 1969 to 1985. The duplication coefficient remain much the same, with perhaps just a very marginal increase in the level of between-channel switching (to an average of 1.1 in 1985). Two larger deviations, the high 2.2 for BBC2 in 1974 and the low 1.5 for BBC1 in 1980, do not appear to be part of any general trends. It is the general steadiness of the findings which is noteworthy.

Table 3.16
Constants of the duplication of viewing law in the UK, 1969–85

Within-channel duplication constants:	1969	1971	1974	1980	1985
ITV × ITV	1.7	1.7	1.8	1.7	
BBC1 × BBC1	1.7	1.8	1.8	1.5	1.6**
BBC2 × BBC2	*	(1.9)	2.2	1.8	
Between-channel duplication constants:					
ITV × BBC1	0.8	0.8	0.9	0.9	
ITV × BBC2	*	0.9	0.9	1.0	1.1**
BBC1 × BBC2	*	(1.1)	1.1	1.2	

Note: *BBC2: not analysed in 1969; adjusted in 1971 (see §3.9)
        **Approximate

Channel 4

Audience duplication between two programmes on Channel 4 (in 1982) was high, with a duplication coefficient of about 4. But this does not mean that Channel 4 programmes have a very high proportion of their audiences in common, since the ratings were generally very low. Thus, if one programme had a rating of 2, then about $4 \times 2 = 8\%$ of the audience of another programme also watched it. But as many as 92% did not.

The high duplication coefficient of 4 only means that duplicated audiences on Channel 4 are high compared with Channel 4 ratings. But during its first year or so, only some 40% of the population watched any of Channel 4's output in a week, compared with 80% for BBC2 and 95% for ITV and BBC1 (see Chapter 5). As for BBC2 in 1971, we can therefore 'explain' the high duplication coefficient by taking into account this low reach. If we re-state Channel 4's ratings as percentages of those viewing the channel at all in a week, they increase by a factor of 2.5. Compared with these adjusted ratings, the duplication coefficient is reduced to about 1.6. This is in line with the within-channel coefficients in Table 3.16 for the other three channels if these are similarly adjusted for their channel's reach.

The duplication coefficients of Channel 4 with the other three channels are in line with the between-channel coefficients for the existing channels. In November/December 1982 the Channel 4 duplication coefficient with BBC1 was 1.0, that with ITV 1.1 (it shares some cross-promotion) and that with BBC2 1.3 (it has most in common with BBC2's programming).

Changes in research methodology

Since late 1984, measurement of individual viewing behaviour in the UK has changed to a 'people-meter' push-button system rather than to the previous method of diary monitored by a set-meter. The definition of 'viewing' has also been changed from a minimum of 8 minutes per quarter-hour to one of effectively 15 seconds to qualify as a viewer. And more guest viewing is being measured. In the event, recorded viewing levels are somewhat higher, maybe by 10 or 20% or more. At the time of writing the audience duplication levels with the new system are not known.

There was also a change in measurement procedures in the UK in the late sixties, with the research contractor changing from Television Audience Measurement (TAM) to AGB. The basic measurement procedures remained the same, but there was round about then a systematic change in duplication coefficients from a relatively low within-channel duplication of 1.3 or 1.4 in 1966 and 1967 (the earliest periods analysed in this way) compared with the post-1969 figures of 1.7 or more, and a marginally higher between-channel duplication of 1.0 in 1966 and 1967 compared with the subsequent 0.8 or so (see Table 3.16). Since 1969 there has been therefore a somewhat more marked (but by no means higher) degree of channel loyalty than before. This is the only sizeable systematic change in duplication levels in the UK observed so far.

3.12 Summary

The number of people viewing two programmes shown on different days of the week generally depends, first, on their overall popularity and, second, on whether they are on the same or different channels. A programme with a rating of 20 will be watched by about – or just under – 20% of the audience of any programme shown on a *different* channel on another day. It will be watched by around 35% of the audience of a programme shown on the *same* channel on another day. Both results are summarised by the duplication of viewing law where the constant reflects the effect of channel loyalty.

4 Other Factors in Audience Overlap

The patterns of channel loyalty and audience duplication described so far have been studied over a wide range of conditions.

In this chapter we first consider some simple generalisations across different demographic groupings (e.g. men and women; geographic regions). Secondly, we discuss more complex situations, namely week-end viewing, off-peak viewing and duplication of viewing between programmes shown on the same day – the 'inheritance effect'. Thirdly, we examine the influence of different types of programme. We conclude with a specific application of the findings to a change in the programming policy for news broadcasts.

Table 4.1
Summary of duplication coefficients for housewives and men

| April 1971 | Duplication-coefficients for | | |
	London Housewives	London Men	Lancashire Housewives
Within ITV	1.7	1.8	1.6
Within BBC1	1.8	1.8	1.6
Within BBC2	1.9	1.8	(1.4)*
Between ITV and BBC1	0.8	0.9	0.9
Between ITV and BBC2	0.9	0.9	(1.0)*
Between BBC1 and BBC2	1.1	1.2	(1.2)*

Note: *Small sample base

4.1 Demographic factors

In general, the duplication of viewing law holds for different demographic groups. Table 4.1 compares the duplication coefficients for London house-wives discussed in the last chapter with those for London men and Lancashire housewives in April 1971. The results are very similar. The within-channel duplications in Lancashire are fractionally lower than in London and there are other small differences, but the broad pattern of results is the same. This basic pattern has also been found to hold across the whole of the UK. And, as we said in Chapter 3, the pattern has generally not changed over time.

4.2 Viewing at off-peak times

Certain exceptions to the simple duplication of viewing law do occur. The major ones are in the afternoon and early evening on the one hand, and late in the evening on the other hand.

Audience duplication between afternoon and early-evening programmes shown on different week-days on a given channel is higher than the audience overlap between programmes shown at peak-times. (This was already foreshadowed by the abnormally high duplications for pairs of programmes at 7 pm noted in the previous chapter.) High duplication also occurs for pairs of programmes shown *late* in the evening on different days.

Table 4.2 illustrates the findings for some late-afternoon programmes on ITV for London housewives in April 1971. Audience overlap is substantially higher than indicated by the duplication law using the 'peak-time' coefficient of 1.7. Of those London housewives who saw 'Lost in Space' at 5 pm on Monday, 37% also watched ITV at 5 pm on Tuesday. This is more than 4 times as high as the 9% of all London housewives who watched then.

Again, 51% of the viewers of 'Lost in Space' watched 'Today' at 6 pm on

Table 4.2
High audience duplication in the late afternoon

	% of Monday Audience who on TUESDAY Viewed ITV	
London Housewives *April 1971*	Junior Showtime 5 pm	Today 6 pm
MONDAY ITV		
Lost in Space – 5 pm	37	51
Today – 6 pm	21	62
Average	25	54
1.7 × Rating	**15**	**34**
Rating	9	20

Tuesday, which is 2.5 times as high as its housewife rating of 20. Not only is the overlap generally higher than at peak-times, but the extent of it also varies from case to case (the overlap usually being highest at exactly the same times on the different days). This is the general finding for afternoon and early-evening viewing on week-days.

Similarly, analysis of late-night viewing in April 1971 shows that of those who watched ITV at 11 pm on the Monday, 39% watched ITV again at the same time on Tuesday. This is three times as high as the 13% rating in the whole population.

The question now is in what ways these early or late off-peak viewers differ from peak-time viewers. First of all, there are fewer of them (the ratings are lower at these times, which is why they are called off-peak). But since so many off-peak viewers watch programmes at similar times on different days, it might appear that they are especially avid viewers of television.

This is really not so. Off-peak viewers do not differ from peak-time viewers in their viewing of *peak-time* programmes. This is illustrated in Table 4.3. The striking finding is that the early-evening viewers on Monday generally behaved just like the viewers of later programmes in terms of the proportion who watched the next day's peak-time programmes. For example, 'The Adventurers' on Tuesday was seen by roughly 50% of both the early-evening and of the peak-time viewers on Monday. The same was true for 'The Saint' at 9 pm and for 'News at Ten'. But 'Bless This House' at 7 pm on Tuesday was on the border line of the 'early evening' effect. It was viewed by a somewhat higher proportion (69%) of the Monday late-afternoon and early-evening viewers than of the peak-time viewers (55%). The pattern shown in Table 4.3 generalises for other days and also for late-night viewers. These also hardly differ in their viewing of peak-time programmes on other days from the general patterns of peak-time viewers themselves.

It follows that the duplication of viewing law as such holds and that off-peak viewers are not unusual in their peak-time viewing. Instead, they merely have a higher tendency than the population at large to view at similar off-peak times on other days. The explanation is that it is the *non-viewers* at off-peak times who are abnormal.

Many people are not at home on week-day afternoons or early evenings, or alternatively do not stay up late in the evening, e.g. past 10 or 11 pm. These things are generally regular social habits – working hours, or one's choice of bedtime, tend to be the same on different days of the week.

The high audience duplication at off-peak times therefore reflects the relatively small pool of available viewers at these times. The ratings are depressed because many people are not available. In comparison with these low ratings, the audience overlap for those viewers who *can* watch is then high.

Table 4.3
The duplication of early and peak-time ITV audiences

London Housewives April 1971		% of Monday Audience who on TUESDAY Viewed			
		Bless This House 7 pm	The Adventurers 8 pm	The Saint 9 pm	News at Ten 10 pm
MONDAY					
Lost in Space	– 5 pm	69	43	46	57
Today	– 6 pm	69	57	49	52
Opportunity Knocks	– 7 pm	69	50	41	45
World in Action	– 8 pm	60	48	43	44
J.G. Reeder	– 9 pm	53	46	46	46
News at Ten	– 10 pm	53	43	39	53
Average		62	48	44	49
1.7 × Rating		**56**	**48**	**39**	**49**
Rating		33	28	23	29

This is similar to the explanation of the apparently high duplications on BBC2 noted in the last chapter (Table 3.12). This was also due to non-availability, in the sense that an appreciable proportion of people were not able to receive BBC2 in 1971. (For the analysis of off-peak viewing no direct quantitative data on the % pattern of viewers' absences from home or their bedtime habits are available, so that no detailed numerical reconciliation of the observed results is yet possible.)

4.3 Week-end viewing

The above interpretation is supported by the findings for week-end viewing when social habits tend to be different. Week-end viewing also follows the duplication of viewing law, as illustrated in Table 4.4 for the overlap between Monday peak-time audiences and those on the following *Sunday*. But the proportionality factor in this case was only 1.3 compared with the 1.7 value for the more general week-day results discussed earlier.

The relatively lower duplication indicates that on Sunday evening the ITV audience was joined by some viewers who were relatively light viewers of the channel during the week. This might be another feature of general social habits – some people have more leisure to view at week-ends. But Table 4.4 applies to April 1971, and at other times the week-end effect has been less marked and the duplication coefficient closer to that for pairs of week-days.

This may be due to seasonal factors affecting social habits, but the general character of the situation is not yet fully established.

Table 4.4
Audience overlap between week-days and week-ends

London Housewives April 1971	% of Monday Audience who on SUNDAY Viewed			
	Stars on Sunday 7 pm	The Ceremony 8 pm	The Ceremony 9 pm	News 10 pm
MONDAY				
Opportunity Knocks – 7 pm	39	62	64	55
World in Action – 8 pm	32	61	62	52
J.G. Reeder – 9 pm	26	52	60	57
News at Ten – 10 pm	25	48	61	61
Average	30	56	62	56
1.3 × Rating	**29**	**52**	**60**	**60**
Rating	22	40	46	46

Table 4.5 extends the picture to week-end afternoon audiences. This reveals higher audience duplication between week-day and week-end afternoon audiences, the proportionality factor being 2.3. This higher duplication is however much less marked than the factors of 3 or 4 seen in Table 4.2 for off-peak duplication on two *week-days*. The reason is that at the week-end the

Table 4.5
AFTERNOON overlap between week-days and week-ends

London Housewives April 1971	% of Monday Audience who on SUNDAY Viewed			
	The Big Match 2 pm	Captain Boycott 4 pm	The Golden Shot 5 pm	H.R. Pufnstuf 6 pm
MONDAY				
Lost in Space – 5 pm	29	46	49	46
Today – 6 pm	21	40	47	35
Average	25	43	48	40
2.3 × Rating	**25**	**46**	**48**	**39**
Rating	11	20	21	17

potential audience is greater than on week-days, thus lowering the degree of duplication from that which occurs on two week-days.

4.4 The inheritance effect

We now turn to how audiences view different programmes shown on the same channel on the same evening. This used to be one of the more misunderstood aspects of audience analysis. Programme planners and television commentators have often subscribed to the belief that once viewers have switched on their sets, they tend to continue watching the same channel throughout the evening. Such a belief puts a high premium on attracting a large audience early in the evening. Once attracted, it is regarded as relatively easy for subsequent programmes on that channel to 'inherit' part of this audience.

However, this common belief is not altogether correct. It has arisen because the within-channel audience overlap for two programmes on the same day is in fact high compared with the ratings of each programme. Table 4.6 illustrates this for peak-time ITV programmes on a Monday night. The ratings are in the low 30s, the duplication in the high 50s and 60s.

Table 4.6
High duplications on the same day

| | % of Monday Audience who on SAME Monday watched ITV | | | |
London Housewives April 1971	Opportunity Knocks 7 pm	World in Action 8 pm	J.G. Reeder 9 pm	News at Ten 10 pm
MONDAY ITV				
Opportunity Knocks – 7 pm	–	65	55	55
World in Action – 8 pm	69	–	69	54
J.G. Reeder – 9 pm	57	69	–	65
News at Ten – 10 pm	55	51	62	–
Monday Rating	32	30	30	32

However, we have already seen that within-channel levels of audience duplication tend to be higher than ratings *even if the programmes are transmitted on different days*. This is the channel loyalty phenomenon isolated in Chapter 3, but its existence had not been generally appreciated. The high degree of audience duplication illustrated in Table 4.6 turns out to be virtually identical to that expected from channel loyalty generally, *except* for pairs of programmes that are shown very close to each other.

Table 4.7
Observed same-day duplications (O), predictions (P) and deviations (O–P)
(predictions based on the between-day duplication law with coefficient 1.7)

	% of Monday Audience who on SAME Monday watched ITV							
London Housewives April 1971	Opportunity Knocks 7 pm		World in Action 8 pm		J.G. Reeder 9 pm		News at Ten 10 pm	
	O	P	O	P	O	P	O	P
MONDAY ITV								
Opportunity Knocks – 7 pm	–		65	51	55	51	55	54
World in Action – 8 pm	69	54	–		69	51	54	54
J.G. Reeder – 9 pm	57	54	69	51	–		65	54
News at Ten – 10 pm	55	54	51	51	62	51	–	
	(O–P)		(O–P)		(O–P)		(O–P)	
Deviations								
Opportunity Knocks – 7 pm	–		14		4		1	
World in Action – 8 pm	15		–		18		0	
J.G. Reeder – 9 pm	3		18		–		11	
News at Ten – 10 pm	1		0		11		–	

This is illustrated by Table 4.7, where predictions from the *between-day* duplication of viewing law are compared with the observed duplication levels between the different ITV programmes on the same day.

These predictions hold to within a few percentage points for the six programme pairs which are separated by more than an hour. The relevant deviations in Table 4.7 are only 4, 3, 1, 1, 0 and (averaging 1.5), as shown in the bottom half of the table where the deviations between the observed and predicted figures are given. More generally, across a wide range of other cases, such deviations have tended to average at zero. The between-day law therefore successfully predicts the within-day duplication pattern for the more widely separated programmes. Generally there is no special overlap for programmes on the same day.

In contrast, the observed overlap is always consistently higher for adjacent or adjacent-but-one pairs of programmes. This is seen well from the bottom half of Table 4.7, the selection of illustrative programmes here being broadcast at hourly intervals. A higher proportion of the viewers of one ITV programme will watch the next ITV programme than if the two programmes were shown on different days altogether. Much the same effect occurs on the BBC channels.

The existence of such 'inheritance' or 'lead in' effects has been known for a long time. Kirsch and Banks (1962) for example noted high correlations in the

USA among the audiences of programmes on the same network on the same evening, and found the highest figures among immediately adjacent programmes. But in the absence of norms for the *usual* amount of audience duplication, neither the size nor the limitations of this effect could be established (e.g. that the correlations for non-adjacent programmes are not specially high).

These then are general findings. It is now clear why many interested parties have overestimated the significance of catching a large audience early in the evening. It is because the basic nature of channel loyalty (as reflected in the duplication of viewing law) was not widely understood. The special inheritance effect for programmes close together is additional to this and can be due to three possible causes. Either people stay tuned to the next programme out of inertia, or because the programme has ended part-way through the programmes on the alternative channels, or because they tuned in to the *previous* programme to wait for the programme scheduled to appear next (a 'lead out' rather than a 'lead in' effect).

4.5 Programme types

Implicit in the findings described so far is the conclusion that while a programme's content undoubtedly affects the size of its audience or 'rating', content has little or no additional effect on the degree of its audience duplication with any other programme. This is a somewhat startling result. Many people would expect that 'clusters' of programmes exist which tend to be viewed by certain groups of viewers. For example, one might suppose that there are viewers who like situation comedies and hence view all or most of them. Alternatively, one might expect a negative, inhibitory effect, e.g. that being addicted to *one* situation comedy or *one* western series is enough – there is no need to watch the other series of the same type.

But in general, neither of these views fits the facts of actual viewing behaviour. Instead, we find that audience overlap for the population at large can be successfully predicted by the duplication of viewing law (plus systematic deviations for early-evening viewing etc. which relate to audience availability). These predictions can be made without taking into account what particular programme or type of programme each audience is watching.

Any programme-type preferences can in effect exist only within the limits of the deviations of the observed duplication figures from the predictions of the duplication of viewing law. These limits have been illustrated in the various tables in this and the preceding chapters. The deviations are generally small, of the order of a few percentage points.

However, some scope for systematic patterns remains. To demonstrate this, the deviations from the simple duplication law are illustrated once more in

Table 4.8 for the Monday and Tuesday ITV programmes discussed earlier. The largest discrepancy, +13 for the two 7 pm programmes, reflects the systematic late-afternoon/early-evening pattern which has already been noted. But the other deviations still range over 12 percentage points, from −5 to +7 (although the average size of the deviations is still only about 3 points). The question now is whether any generalisable patterns are reflected by the relatively low duplication between 'World in Action' and Tuesday's 'News at Ten' say, or by the high duplication between 'J.G. Reeder' and 'The Saint'.

Table 4.8
Observed duplications (O), predictions (P) and deviations (O–P)
(from Tables 3.3 and 3.5)

London Housewives April 1971	% of Monday Audience who on TUESDAY Viewed							
	Bless This House 7 pm		The Adventurers 8 pm		The Saint 9 pm		News at Ten 10 pm	
	O	P	O	P	O	P	O	P
MONDAY								
Opportunity Knocks – 7 pm	69	56	50	48	41	39	45	49
World in Action – 8 pm	60	56	48	48	43	39	44	49
J.G. Reeder – 9 pm	53	56	46	48	46	39	46	49
News at Ten – 10 pm	53	56	43	48	39	39	53	49
	(O–P)		(O–P)		(O–P)		(O–P)	
Deviations								
Opportunity Knocks – 7 pm	13		2		2		−4	
World in Action – 8 pm	4		0		4		−5	
J.G. Reeder – 9 pm	−3		−2		7		−3	
News at Ten – 10 pm	−3		−5		0		4	

Virtually no significant patterns in these deviations have, however, been found over the years for any particular type of programme. Table 4.9 illustrates the first such analysis, for data for the first week in May 1967. It sets out 20 programme categories and sub-classifications for ITV and BBC1 programmes (excluding children's programmes), together with the number of programmes in each category and their average ratings. The last column gives the average deviations of the observed audience duplication from the predictions of the duplication of viewing law for each pair of programmes within the category. In analysing such aggregated data it is more convenient to express the readings as percentages of the population seeing both programmes of a pair, rather than as percentages of the audience of one also seeing the other. (See also the Appendix of Chapter 2 for this formulation of the duplication of viewing law.)

The average deviations are generally less than 1 rating point, which is small

Table 4.9
Deviations of the observed duplications from the duplication of viewing law analysed by
20 standard IBA programme categories and classes
(pairs of programmes on different evenings but excluding programme pairs near 6 pm;
ITV & BBC1, 1–7 May 1967, London & North housewives)

CATEGORY & Class	Number of programmes	Average rating	Av. observed minus theoretical duplications*
NEWS (Week-day)	40	23	−.3
NEWS MAGAZINES	22	21	.6
DOCUMENTARIES & NEWS FEATURES			
News features	19	15	.4
General discussion	11	15	.5
Magazines	4	16	.5
Arts	9	12	−.4
Miscellaneous	12	15	.1
RELIGION	20	7	.4
ADULT EDUCATION	7	5	.4
PLAYS	8	20	.7
DRAMA & SERIALS			
Series	9	26	−.2
Adventure & crime	29	28	.1
Westerns	6	25	1.2
Serials	12	28	.0
CINEMA FILMS	14	32	−.1
ENTERT. & MUSIC			
Comedy series	24	23	.1
Contests	12	26	.7
Light music	9	19	.1
Other	11	22	.1
SPORT	25	15	.8
AVERAGE	15	21	.2

Note: *Duplications expressed as per cent of the population

compared with the audience rating of each programme. In effect therefore, there are therefore virtually no systematic differences from the law's predictions and no evidence of clusters by programme type.

'Westerns' provide the only programme grouping for which the average deviation is more than 1 rating point. To this extent there was a special tendency for people to watch two or more of the westerns as a group. However, this result is based on only six programmes, is of doubtful statistical significance, has not been found in later years, and is small anyway. It seems therefore at most the kind of (small) exception to prove the rule that, as far as the IBA's commonsense programme classifications are concerned, there is no special tendency across the population for people who watch *one* programme of a given type also to watch others of the same type.

Checks among *regular* viewers of programmes (i.e. those who see at least three episodes out of four) have also shown no evidence so far of any clustering in viewing behaviour by programme type for them. Nor does clustering show itself in one-person households, where individual viewing preferences could operate irrespective of conflicting preferences of different family members.

Some writers in the US, e.g. Swanson (1967), Bruno (1973), Gensch and Ranganathan (1974) and Headen et al (1979), have reported apparent programme-category effects derived from applying statistical techniques like factor analysis or multiple regression to certain types of viewing data. But Swanson, for example, did not take account of the choice of channel (or day of week) and failed to note that all the programmes in his first factor were transmitted by the ABC network, all those in the second factor by CBS and so on (Ehrenberg 1968). The high correlations reflected by these so-called programme factors were therefore no more than a reflection of channel loyalty. The other studies cited are also open to interpretative difficulties, including differences between people's *reported* viewing behaviour (as discussed in Chapter 9) and directly measured behaviour.

4.6 A specific application: news broadcasts

A practical application of the general results outlined in this and the preceding chapters is illustrated by an evaluation of the audience for news broadcasts. One specific case arose some years ago when certain decisions had to be made regarding ITV's news bulletins. Until July 1967 both of the main TV channels in Britain carried two 10-minute news bulletins per day, screened on week-days at about 6 and 9 pm.

In July 1967 the later ITV bulletin was increased to 30 minutes to allow greater depth of treatment and was also moved from 9 to 10 pm. This involved three changes, the time of the bulletin, its length and the fact that it was no longer opposite the 9 pm bulletin on BBC1. (BBC2 was not yet widely viewed at the time.) To evaluate the situation, studies were made in May 1967 (well before the change), just after the change in July, and in October when viewing behaviour appeared to have fully settled down again.

In trying to assess the public's response to the flow of news, one might have measured how well major news items had been communicated (in May 1967 these included violence in Aden, Britain making an application to join the European Common Market and hopes for a quieter jet engine). But it was not clear how any assessment of communication (by currently feasible research techniques) could tell one much about the effectiveness of TV news bulletins in general and the effect of the July programme change in particular.

A more direct, if limited, evaluation seemed to be in terms of people's actual utilisation of the news bulletins, namely in terms of their viewing behaviour. How many people did not see *any* news? At the other extreme, was there any special following for TV news and, if so, was this increased or decreased by the change to a later and longer ITV bulletin?

The ratings tell us the audience size achieved by each separate news bulletin, but a further question is what use people make of the *different* bulletins available to them. In what combinations do people view bulletins on the same day and on different days, on the same channel and on different channels, at 6 pm and at 9 or 10 pm? Given that the contents of the different bulletins on the same evening would usually not vary greatly, did viewing of one bulletin per evening suffice and therefore inhibit the watching of any others?

To try to answer such questions it is not enough merely to tabulate viewing patterns. One has also to be able to predict what such patterns would generally be like. Only if the factors involved are sufficiently well understood to permit successful prediction can the effect of variations in any of the associated factors (such as the ITV programming change) be evaluated.

The basic step therefore has to be to compare the observed viewing patterns for the news with the predictions of the duplication of viewing law as an interpretative norm. In 1967 the coefficients for this law, for the general run of programmes on different week-days on the two channels, were 1.4 for within-channel duplication on both ITV and BBC1 and 1.0 for between-channel duplication.

The news on different days

Table 4.10 shows the extent to which viewers of one news bulletin watched others on another week-day, either two bulletins on the same channel (e.g. ITV × ITV or BBC × BBC) or on two different channels (ITV × BBC).

The results here are again set out in terms of the percentage of the total population viewing both programmes (rather than in the format used earlier, showing the percentage of the audience of one programme who also watched the other). Thus in May, prior to the programming change, on average 9% of the housewife population watched both the 6 pm ITV bulletins on two different days (the first line of figures). Similarly, 9% of the population watched a 6 pm ITV bulletin on one day and a 9 pm ITV bulletin on the other. (This is just over half the audience of the average 6 pm bulletin with a rating of 17, but only a quarter of the larger audience at 9 pm with a rating of 36. Expressing the

Table 4.10
*Observed duplications (O), predictions (P) and deviations (O–P) for news
bulletins on different week-days*
(averages across the five week-days – 1–5 May 1967)

Housewives London & the North	Observed Ratings	% of Population Viewing Both Bulletins		
		O	P	(O–P)
ITV × ITV				
6 pm × 6 pm	17 & 17	9	**4**	5
6 pm × 9 pm	17 & 36	9	**9**	0
9 pm × 9 pm	36 & 36	18	**18**	0
BBC × BBC				
6 pm × 6 pm	15 & 15	6	**3**	3
6 pm × 9 pm	15 & 24	5	**5**	0
9 pm × 9 pm	24 & 24	8	**8**	0
ITV × BBC				
6 pm × 6 pm	17 & 15	3	**3**	0
6 pm × 9 pm	17 & 24	3	**4**	−1
9 pm × 6 pm	36 & 15	5	**5**	0
9 pm × 9 pm	36 & 24	8	**9**	−1

individual audience duplications in such *relative* terms is often more informative, but it is less succinct when analysing extensive data as here.)

Comparisons with the predicted duplications in Table 4.10 gives clear-cut results. Duplication of viewing for news bulletins on different days is mostly the same as for programmes generally, i.e. it is as predicted by the duplication law.

There are only two apparent exceptions – the high duplications for two 6 pm ITV or two 6 pm BBC bulletins. But even these deviations are predictable. They are part of the high duplications in the afternoon and early evening which occur generally, as described earlier in this chapter. The evidence is that if other programmes were screened at 6 pm on two days, they would similarly tend to draw an apparently abnormally high number of common viewers on different evenings.

News bulletins on the same day

Next we turn to the 6 and 9 pm bulletins on the same day. These times are far enough apart for the duplication law also to hold without any special 'inheritance effect', such as occurs for more or less adjacent programmes on the same day (Table 4.7). Nonetheless, the similarity of content of the different bulletins might be expected to lead to abnormal effects.

Table 4.11 shows that this is not so: the same-day viewing patterns were in fact virtually normal. Thus, with two ITV programmes rated 19 and 36 and a duplication coefficient of 1.4, one would expect 10% of the population to watch both ($1.4 \times 19 \times 36/100 = 10\%$). That is what occurred.

Table 4.11
Same-day duplications for news bulletins

Housewives London & the North May 1967		*Observed Ratings*	*% of Population Viewing Both Bulletins*		
			O	P	(O–P)
6 pm ITV	**9 pm** ITV	19 & 36	10	**10**	0
ITV	BBC	19 & 25	5	5	0
BBC	ITV	15 & 36	4	5	−1
BBC	BBC	15 & 25	5	5	1

It follows from these findings that there was no inhibition about watching more than one news bulletin. In fact, the above results imply that of people who watched a 6 pm bulletin (ITV or BBC), about 75% would also watch one at 9pm. This result is neither specially 'high' or 'low' – it is just about what would be found for any programmes in these time-slots with the corresponding rating levels. Similarly, about 40% of the audience of a 9 pm bulletin would already have seen a 6 pm one. (The ratings at 6 pm are lower than at 9 pm, as shown in Table 4.11, and hence the duplicated viewers are a lower proportion of the 9 pm audience than of that at 6 pm.)

'News at Ten'

These various results implied that putting on a half-hour ITV news bulletin at 10 pm instead of a shorter one at about 9 pm need lead to no peculiar results. And so it turned out.

Table 4.12 shows that in October 1967, for instance, the percentage of viewers of 'News at Ten' among viewers of other news bulletins was close to the predicted levels. A typical 'News at Ten' was watched by about the predicted 46% of audiences of other ITV bulletins and by just under the predicted 31% of the audience of BBC bulletins.

A special result concerns the audience duplication between the BBC 9 pm News and the ITV 'News at Ten' on the same day. Previously, when both

Table 4.12
Duplication of 'News at Ten' with other news: observed (O) and predicted (P)

Adults in London October 1967		% of other News Audience Who Viewed News at Ten	
		O	P
ITV	News at 6 pm, Same Day	45	46
ITV	News at 6 pm, Diff. Day	43	46
ITV	News at 10 pm, Diff. Day	47	46
BBC	News at 6 pm, Same Day	28	31
BBC	News at 6 pm, Diff. Day	28	31
BBC	News at 9 pm, Same Day	27	31
BBC	News at 9 pm, Diff. Day	30	31

channels had screened news bulletins at the same time at 9 pm or within a few minutes of each other, watching both was either impossible or required very deliberate switching (which could not in effect be measured by recording of viewing in quarter-hour time-periods). But as the table shows, after the scheduling change, about 27% of the audience of the 9 pm BBC bulletin also watched ITV's 'News at Ten' – almost the same as if these were *any* kinds of programmes.

Complementary programming

ITV's change to 'News at Ten' occurred with virtually no effect on rating levels as such. But one consequence of the duplication results was that slightly fewer people saw at least one news bulletin per day. Previously, with 'competitive' programming, somebody watching at 9 pm had to see the news on one channel or the other (few people were able to receive BBC2 in 1967). Now, with 'complementary' programming on BBC1 and ITV at both 9 and 10 pm, people need not watch the news at *either* time since they were provided with the choice of a non-news programme on the other channel.

The total number seeing any news therefore drops. This is balanced, in arithmetical terms, by the increased number of people who see both bulletins (and indeed, a *longer* one at 10 pm). But the drop in the total numbers seeing news does not imply any special tendency to avoid the news. Similarly, the fact that some people watch both the 9 pm BBC and the 10 pm ITV news does not imply that they are specially avid followers of the news. Virtually the same duplication percentages would have occurred for any programmes with the same rating levels.

4.7 Summary

The duplication of viewing law holds for different demographic groups and regions of the country.

Duplication of viewing between different week-day afternoons or early evenings is relatively higher. This appears to be due to the consistent non-availability of those people who are at work rather than due to any specially intensive viewing by those who actually view. The same high duplication pattern occurs consistently between late-evening programmes on different days, due to fairly consistent bedtime habits by part of the population.

Consecutive or near-consecutive programmes on the same evening share their audience to an above-normal extent, but this 'audience inheritance' does not extend to programmes further apart. This appears to rule out the possibility that a large channel audience can be held throughout an evening by transmitting a very popular programme at the beginning of peak-time. It does, however, leave room for the strategy of 'hammocking': boosting the audience to a programme by placing it between *two* very popular programmes.

There are no special duplication patterns for programmes of any particular type (e.g. comedies, current affairs, etc.). As far as people's actual viewing behaviour is concerned, different programmes of the same type do not appeal especially to the same viewers.

5 Repeat-Viewing

We now turn to the subject of repeat-viewing, i.e. the extent of audience overlap between different episodes of regular programmes, usually screened one week apart. The basic questions are how many viewers of one episode will also watch the next, and on what factors this audience overlap depends, e.g. the type of programme, its popularity or rating, the time of day, etc.

The overall result is simple if perhaps surprising. For popular programmes only around one half – generally some 55% – of the people who see a regular programme one week also see the next episode of it in the following week. This was first noted in the UK in the late sixties. But it still remains true 20 years later and it also holds in the USA and other countries. The growth from two channels to four in the UK and the beginnings of the current 'media revolution' (VCRs, Cable, etc) have not led to markedly less programme loyalty despite more opportunities to choose, nor yet to *more* loyalty in order to avoid the need to choose. This is probably because the relatively low repeat levels arise chiefly from a lack of regularity in viewing television *at all* at a given time from one week to the next, rather than from switching to a different programme.

More recent analyses have also brought out a link between repeat-viewing and overall popularity. Programmes with low ratings have repeat-viewing levels below even the average level of about 55%, and often well below. These results, which also reflect on less popular channels and less popular programme types, are set out in Section 5.5. They show that small rating programmes are not generally watched by viewers who are exceptionally loyal to them.

5.1 'Z-Cars'

We begin with an historical illustration for a particular programme. The popular weekly police programme 'Z-Cars', popular in the UK in the sixties and seventies, is typical of many TV series: episodes with individual stories but involving the same main characters are shown at the same time each week, for

'Z-Cars' on BBC1 on Monday night from 7–7.45 pm. Sometimes, however, (e.g. in 1967) a slightly longer script was used and screened in two separate halves each week on Mondays and Tuesdays from 7–7.30 pm.

In a typical week 24% of London housewives then watched the first half of the programme and 24% also watched the second half, as shown in Table 5.1. One might think that these would be the same people, but that was certainly not the case.

Table 5.1
The audience for the two halves of 'Z-Cars'

London Housewives 1967	% Viewing
1st half, Monday 1 May	24
2nd half, Tuesday 2 May	24

In fact, 12% of the population analysed watched both halves of 'Z-Cars', as set out in Table 5.2. Only 12 out of 24, or 50% of those who watched the first half of the programme on Monday, also watched – or should one say 'bothered to watch' – the second half on Tuesday. And correspondingly, of those who watched on Tuesday, as many as 12/24 or 50% did so without having seen the beginning of the episode the previous day. Yet this turned out to be typical also for repeat-viewing of different episodes altogether.

Table 5.2
Viewers of BOTH halves of 'Z-Cars'

London Housewives 1967	% Viewing
Both Monday and Tuesday 7–7.30 pm	12

Such relatively low audience overlap between two parts of a programme is however higher than if the broadcasts had been of two different programmes, i.e. higher than the level of audience duplication discussed in earlier chapters. Thus the duplication coefficient for BBC1 programmes in general in 1967 was about 1.4. For two programmes with ratings of 24 as 'Z-Cars', the predicted audience duplication would be about $1.4 \times 24 \times 24/100 = 8$; i.e. a third of the audience should be in common. But such a prediction of course applies only to two different programmes, not two episodes of the same series, let alone two halves of the same episode as here. One might well expect that for a programme like 'Z-Cars' the pull of watching the concluding half of a gripping episode the next day would be much higher. In the event it was higher – there is some programme loyalty – although apparently not that much.

5.2 55% repeat-viewing

Table 5.3 gives a more general illustration, namely week-by-week repeat-viewing results for 40 regularly screened programmes (including film slots and the like) in the spring of 1971, arranged in decreasing order of their rating levels. The striking finding is that generally the percentage of viewers of one episode who watched the next one in the following week is roughly 55%. This holds

Table 5.3
The percentage of the audience viewing again the following week
(London housewives, April/May 1971)

	Average Rating	Repeat %		Average Rating	Repeat %
Doctor at Large	35	56	This Week	25	45
Coronation Street	34	63	Bev. Hillbillies	24	62
Sunday Film	34	50	Sportsnight	24	53
This is Your Life	34	68	Cinema	23	49
Dick Emery	32	59	The Virginian	22	60
Ironside	31	58	Val Doonican	22	46
Nearest & Dearest	30	58	Top of the Pops	21	56
Thursday Film	30	58	Golden Shot	21	51
Tuesday Film	30	58	FBI	20	49
Smith Family	30	58	Stars on Sunday	20	49
Two Ronnies	29	50	Top of the Form	20	53
The Western	29	53	Thursday Play	20	42
Opportunity Knocks	29	66	Seven Men	19	49
Budgie	28	65	Please Sir	18	44
Hawaii Five-O	28	65	The Doctors	18	61
The Saint	28	64	Coppers End	17	39
Hine	27	55	Name of the Game	17	50
Persuasion	26	44	The Avengers	15	50
Saturday Film	26	44	Braden's Week	16	48
Match of the Day	25	42	Peyton Place	10	51

AVERAGE RATING: 25   AVERAGE REPEAT %: 54

almost equally for the more popular programmes in the left hand column and for the programmes with somewhat lower ratings shown in the right hand column.

There is some variation in the individual repeat-viewing percentages but it is not large, mostly from about 45% to 65%, with a few exceptional values. To a first order of approximation, for any of the programmes something like *half* the audience watches the next episode a week later.

This is a very simple finding. It is also perhaps a somewhat remarkable one, in two ways. First, that repeat-viewing for this supposedly 'compulsive' medium is no higher; second, that there appears to be so little dramatic variation between programmes. It is highly uncommon for one programme to attract a vastly more loyal following than another. ('Jewel in the Crown', the renowned UK serial of 1984, typically had a repeat-rate from episode to episode of 53%.) There is a downward trend with rating levels but there appear to be few other systematic factors at work, as we now show.

Programme types

Table 5.4 illustrates that there is little if any systematic difference in repeat-viewing levels for different types of programme. Most of the results average at about 55%.

Plays had the lowest repeat level (an average of 42%) in this particular analysis but that is not a general result. In other years the repeat level for different plays in a regular drama slot (e.g. 'The Monday Play') has also been over 50% (e.g. Table 5.6).

Table 5.4
Average repeat-viewing for different programme types

May 1971	London Men	London H/Ws	Lancs H/Ws	**Average**
Serials	51	59	62	**57**
Series	53	57	58	**56**
Comedy series	48	51	58	**53**
Shows	48	54	55	**52**
Quizzes & games	58	57	55	**57**
Sport	47	53	43	**48**
Films	55	56	52	**54**
Plays	35	46	44	**42**
Miscellaneous	52	52	53	**52**
AVERAGE	51	55	55	**54**

Demographic factors

Repeat-viewing levels have not been found to vary greatly for different demographic groups such as men or women, or different regions of the country, as is illustrated in Table 5.4 above and in Table 5.5. A small-scale study of viewing by children aged 4–15 has also revealed an average repeat-viewing level of about 50%.

Week-end viewing

There is some suggestion that in 1971 repeat-viewing levels for regular week-end programmes were fractionally lower than for week-day programmes, as is illustrated in Table 5.5. The difference is not large, but a possible explanation may be in terms of availability: people may be somewhat less settled in their general social habits at week-ends. However, more empirical checks are needed here.

Table 5.5
Repeat-viewing week-ends and week-days

May 1971	London Men	London H/Ws	Lancs H/Ws	**Average**
Week-end	47	44	52	**49**
Week-day	55	59	58	**57**

Rating levels

Perhaps the most important special factor found is a general tendency for repeat-viewing to decrease with rating levels. This can be summarised for the programmes in Table 5.3 by the averages for the five programmes with the highest ratings and the five with the lowest:

	Av. rating	Av. repeat
The 5 highest-rating programmes	34	59
The 5 lowest-rating programmes	15	48

For programmes differing on average by almost 20 rating points, the repeat-viewing percentage therefore dropped by just over 10 points.

This trend is documented further in Section 5.5. The pattern is an example of McPhee's 'Law of Double Jeopardy', that the fewer the people who choose an item (a low-rating programme), the less those who choose it 'like' it (i.e. low repeat). There are many other situations where this law applies (e.g. McPhee, 1963; Shuchman, 1968; Ehrenberg, 1972; Barwise and Ehrenberg, 1984, 1985, 1986; Ehrenberg *et al.*, 1986). Here it means that programmes with small audiences are *not* generally watched by people who are particularly devoted or loyal to them, but the reverse. Low-rating programmes have low repeat-viewing levels.

Non-stationarity

The analysis of repeat-viewing is basically simple when the rating levels of the two episodes are the same. The two halves of 'Z-Cars' analysed earlier provide an example. The rating of each half-episode was 24 and since 12% of the population watched both halves, 12/24 or 50% of the audience of either half also watched the other.

But audience sizes for different episodes of a programme need not be the same. This can obviously happen when one is considering a rather broad type of programme such as drama or film slots where the popularity of individual plays or films can vary markedly. But even with a series or a serial there can be variations in audience levels.

A relatively complex illustration was for the twice-weekly serial 'Coronation Street' back in April 1971. Monday and Wednesday ratings of 'Coronation Street' over two weeks among Lancashire housewives were:

	Week 1		*Week 2*	
	Mon.	Wed.	Mon.	Wed.
	34	39	40	28

The first Monday was Easter Monday, a public holiday with unusual competitive programming and the 'Coronation Street' rating was lower than on the following Wednesday. But in the next week the Monday rating was substantially higher than that on Wednesday, a pattern which also held up in the following weeks and was no doubt due to popular Wednesday programmes on BBC1.

It follows that whereas in Week 1 in April 1971 all the Monday viewers (100%) could have watched the Wednesday episode as well, in Week 2 at most 70% (28/40) of the Monday audience could have done so. These are upper limits to the possible level of repeat-viewing and they affect the repeat-viewing levels actually attained. One would expect Monday to Wednesday repeat-viewing to be lower in Week 2 (and it was – about 40% compared with 60% in Week 1, or with the 63% *week-by-week* repeat-viewing for 'Coronation Street' at the time – see Table 5.3).

In general, such variability in audience size from one episode of a programme to another (which might partly be due to sampling errors in the ratings data) can confuse the analysis and interpretation of repeat-viewing levels. A shared audience of a given size will represent a larger percentage of the smaller audience.

For high-rating programmes these effects will not be great. Thus for two episodes with ratings of, say 31 and 28, a shared audience of 16 points will represent 52% of the audience to the first episode and 57% of the audience to the second. But for two low-rating episodes with ratings of about 6 and 3 the same size difference of 3 ratings points will have marked effects. A repeat-viewing level calculated on the base of the smaller audience ($2/3 = 66\%$) will be twice as large as one calculated on the base of the larger audience ($2/6 = 33\%$). Some care has to be taken therefore in any discussion of repeat-viewing levels for low-rating programmes. In practice, we usually define repeat-viewing using the lower rating as the base.

5.3 Viewing other channels

If only about 55% of the audience of a programme usually watch the next episode of the programme, what are the remainder of the first audience then doing? Do they still watch television but on another channel (possibly because of the conflicting preferences of other family members)? Or do they not watch television at all (and is this more or less unavoidable, like being out, or only due to not feeling like it)?

The main answer is that for fairly popular programmes (ratings of 10 or more) generally only a small percentage of the 'lapsed viewers' watch television on another channel at that time. Table 5.6 illustrates the pattern for the two main channels in June 1969. (At that time the third UK channel, BBC2, was received and viewed only to a very small extent.) In these data about 50% tended to be repeat-viewers and only about 10% watched the other channel instead. Thus, the main factor in *non-repeat-viewing* was that some 40% of those watching a programme in one week did not watch television at all at the same time in the next week. The implication is that if and when people *do* watch again, most watch the same programme as before. Some programme loyalty exists but it is often not compelling enough to make people watch TV specially at that time.

More recent analyses for *low-rating* programmes also show some reduction in programme loyalty, but again not much. For example, low repeat-viewing of programmes on Channel 4 – where ratings tend to be low – is associated with a quite substantial tendency to watch another channel instead in the following week. It remains true, however, that about 40% of the viewers of one episode

are not watching television *at all* at the same time the following week. Viewers
of a low-rating programme are somewhat more likely to watch it again the week
after than non-viewers, but neither are *that* likely to do so. Both the ratings and
the repeat-levels are low.

Table 5.6
Viewing the same or the other channel a week later
(ITV and BBC1)

| London Housewives June 1969 | % of audience who exactly a week later watch | | |
	SAME Channel	OTHER Channel	EITHER Channel
Programme-type			
Twice-weekly serials	62	8	70
Other serials	55	5	60
Series	55	12	67
News at Ten	55	11	66
Documentaries	48	14	62
Quizzes and games	57	6	63
Variety shows	52	13	65
Magazine programmes	51	11	62
Plays	51	9	60
Films	52	16	68
Low-rating programmes	32	12	44
AVERAGE	52	10	62

5.4 Non-consecutive episodes

The finding that for programmes with sizeable ratings non-repeat-viewers
mostly have not switched to another channel suggests that repeat-viewing in
successive weeks is influenced by general social habits and availability which
determine whether one is watching television at all. This in turn should imply
that those viewers who do not watch the next episode of a programme are not in
fact 'lost for good'. Their behaviour is less a reaction against the particular
programme than a reflection of somewhat irregular viewing habits generally.
They may well watch the programme again the week after that.

This is borne out by the facts. There is little if any erosion in the percentage of
repeat-viewers for episodes further than a week apart. Table 5.7 illustrates this

for programmes screened regularly (at the same time each week) over the four-week period in April/May 1971 which we have been discussing.

The average figures stay at over 50%. The slight decline in weeks 3 and 4 is at least in part due to a seasonal decline in rating levels from Week 1 onwards.

Table 5.7
Repeat-viewing in four successive weeks for regular programmes
(regular programmes, April/May 1971)

% of 1st Week audience who viewed in	London Men	London H/Ws	Lancs H/Ws	**Average**
second week	50	55	56	**54**
third week	49	50	55	**51**
fourth week	49	52	52	**51**
AVERAGE	50	52	54	**52**

More recent work has extended this kind of analysis to time-periods longer than the four weeks illustrated here. This shows that even over as many as 10 episodes of regular weekly or daily programmes (e.g. news) audience overlap declines only very slowly as the interval between the pair of episodes is increased. The decline is typically only of the order of one percentage point for each week added to the interval.

5.5 Repeat-viewing and low rating levels

We have briefly referred above to a link, already apparent in the seventies, between repeat-viewing and rating levels. There was a tendency, seen in Table 5.3, for repeat-viewing levels to be lower than 55% for somewhat less popular programmes. Analyses of viewing behaviour in the eighties have covered a wider range of rating levels, clarifying this link between repeat-viewing and overall popularity.

Repeat-viewing levels for 80 programmes screened in the spring of 1980 are summarised in Table 5.8. The programmes are grouped in terms of their audience size, from eight programmes with ratings of under 5 through to seven programmes with ratings of 30 and over. The trend is clear: repeat-viewing levels are broadly in the range of about 45 to 65 noted earlier for fairly popular programmes (ratings of 10 or more) and even in the high 60s for the most popular programmes. But they are 40 or less for low-rating ones. Ignoring the smallest ratings, the repeat-viewing percentage increases very roughly by

about 1 point for a 1-point difference in the rating. For ratings varying by about
25 points from 5 to over 30, the average repeat level increases also by about
25 from 40 to 67.

Table 5.8
Repeat-viewing and rating levels

London and Lancashire 1980	Rating							
	0–	5–	10–	15–	20–	25–	30–	*All*
Number of programmes	8	4	12	22	21	6	7	80
Average repeat %	15	40	42	51	56	67	67	50

Channels

The tendency for repeat-viewing levels to be lower for less popular
programmes must be expected to affect comparisons between channels. This
can be seen in Table 5.9, comparing repeat-viewing levels for programmes on
the four UK channels operating in early 1983.

Table 5.9
Repeat-viewing on the four channels

London 1983	Number of programmes	Average rating	Average week-to-week repeat-viewing
ITV	120	18	49%
BBC1	172	17	47%
BBC2	36	8	34%
Ch.4	156	2.5	34%

The findings for the two major channels agree broadly with the results seen
earlier. With average ratings of 17 or 18, average repeat-viewing levels are just
under 50%. This is a little lower than the 55% 'norm' but only in line with the
decline in the average programme ratings reported at the time.

For BBC2 and Channel 4 repeat-viewing levels are a good deal lower,
averaging only about 35%. Average ratings on these channels were also much
lower, about 8 for BBC2 and only 2.5 for Channel 4. Differences in
repeat-viewing levels between the channels are thus broadly in line with the
relationship between repeat-viewing and ratings seen above. With such low

average ratings, repeat-viewing on Channel 4 at 34% may even be considered quite high.

Viewing other channels

In Table 5.6 we showed that in 1969 most (38 out of 48) of those who failed to repeat-view the next episode of a programme were not then viewing television at all. At that time most viewers had a choice between only two channels: ITV and BBC1. By 1983 the choice had increased for most people in the UK to four channels with the growth of BBC2 and the introduction of Channel 4, plus some VCR ownership already. This might be expected to lead to more conflict in programme choice and hence a greater tendency for people to watch a different channel next week instead of repeat-viewing the same programme.

Table 5.10
Viewing the same or a different programme

London 1983	Number of programmes	Average rating	*Average % viewing in the following week*		
			The same prog.	Another prog.	ANY TV
ITV	120	18	49	13	62
BBC1	172	17	47	13	60
BBC2	36	8	34	24	58
Ch.4	156	2.5	34	24	58

In fact, Table 5.10 shows (from an analysis which centred on Channel 4) that there had been little change in the pattern for the two major channels. Of the audience to an episode of a regular programme on ITV or BBC1 almost 50% watched the same programme the next week (as already just noted), only just over 10% watched a different channel and some 40% did not watch TV then at all. Still, two-thirds of those who did not repeat-view were not watching television at all.

For BBC2 and Channel 4 the picture was different. In line with the lower ratings, repeat-viewing from one week to the next was lower on average (34%), but almost as many Week 1 viewers (24%) watched programmes on *another* channel. In total, viewers of programmes on BBC2 or Channel 4 were not much less likely to view television *at all* at the same time in the following week. (On average 58% did, compared with 60–62% for ITV and BBC1.) These lower-rating programmes tend therefore to attract irregular viewers of the *programme* but not less regular viewers of *television* as a whole. Nonetheless, the average repeat-viewing levels of 34% are much higher than the average programmes' ratings of 8 and 2.5, showing again a degree of repeat-loyalty.

Programme types

Just as the relationship between ratings and repeat-viewing levels affects comparisons between channels, so it affects comparisons between programme types. More serious and demanding information programmes tend to get lower ratings. The findings are that they also attract somewhat lower levels of repeat-viewing from week to week, but only in line with those lower ratings.

5.6 Audience cumulation

The implication of the repeat-viewing results discussed so far is that those people who watch a regular programme do not generally watch every episode of it. The question then arises as to what happens over a longer sequence of broadcasts, especially when the programme is a serial with a continuous story-line over a limited number of episodes.

As an example we consider 'Brideshead Revisited', a serial screened in 11 episodes in the latter part of 1981. The programme received great critical acclaim and the impression was conveyed in the press that it had 'hooked' the viewing public week by week to an exceptional extent. This was not so.

The first episode was seen by about 27% of adults and thereafter about 20% saw each episode. Repeat-viewing from week to week averaged 52%. This is about the normal level for a peak-time programme with a good rating. The question now is how this affected viewing of the serial as a whole?

Table 5.11 shows that with 40% of adults seeing none of the 11 episodes the 60% who did see the programme at least once were made up as follows: 17% saw just one episode, 11% saw two episodes and so on, with only about 4% of

Table 5.11
Number of episodes seen of 'Brideshead Revisited'

London 1981	Number of episodes seen											Av. number seen
	0	1	2	3	4	5	6	7	8	9	10	11	
% of adults	40	17	11	8	4	3	3	4	2	4	2	2	4.4*

Note: *By those (60%) seeing at least one episode

adults actually seeing 10 or 11 of the episodes. The average person who watched the serial at all saw only 4.4 of the 11 episodes. 'Brideshead' was a popular programme but it did not create a new breed of addicted viewer (e.g. only some 4% of the population saw every or nearly every episode).

Reach and frequency

These results for 'Brideshead Revisited' are known to be typical. Relatively few viewers see all or nearly all the episodes in any series.

As an older illustration we consider four successive episodes of 'This is Your Life' in the four weeks ending 9 May 1971. The ratings amongst London men were about 32, with small irregular variations of about one percentage point from week to week. On average, 69% of the viewers of one programme also saw another, whether in an adjacent week or not. (This was a rather high repeat-level even for a high-rating programme.)

Table 5.12 shows that 50% of men saw none of the four episodes, 15% saw one of the episodes, 9% saw two, 13% saw three, and another 13% saw all four. Only just over one in four of those watching the programme at all in the four weeks watched all four episodes, i.e. $13/50 = 26\%$. (We note that the rating for the average or typical episode is made up of a quarter of those seeing only one episode, i.e. 15/4, plus half those seeing two out of four, i.e. 9/2, etc., giving $4.7 + 4.5 + 9.8 + 13.0 = 32$.)

Table 5.12
The distribution of exposures to 'This is Your Life'

London Men April/May 1971	% Seeing the Programme				
	0	Once	2 times	3 times	4 times
All Men = 100%	50	15	9	13	13

Two conventional concepts in dealing with the frequency with which such a 'schedule' of television broadcasts is seen are *reach* and *average frequency* (also commonly used in evaluating advertising campaigns which are however usually scheduled over a mixture of programmes).

Reach is the percentage (or number) of the population who are reached at all by the schedule, i.e. those who see at least one of the broadcasts in question. (The reach of a channel will be similarly defined in the next chapter as the percentage watching the channel at all during a given period.) The reach of 'This is Your Life' over four episodes was therefore 50%. Thus, although about 32% of men saw any particular broadcast, viewers of different episodes were not always the same people and hence a total of 50% saw at least one.

Average frequency is the average number of episodes seen by those reached at all. If people watching one episode also saw all the other three, the average frequency would clearly be four. In our example the total number of episodes seen is built up from the 15% who saw one, the 9% who saw two and so on, giving

$$15 \times 1 + 9 \times 2 + 13 \times 3 + 13 \times 4$$
$$= 124 \text{ episodes}$$

seen per 100 London men. But since only 50% of London men saw *any* of the four broadcasts, the average frequency with which they did so was

$$124/50$$
$$=2.5$$

Thus, half the population saw 'This is Your Life' in the four weeks (the 'reach') and they each saw on average 2.5 out of the four episodes.

The higher the repeat-viewing percentage the lower is the reach and the higher the average frequency of exposure per viewer (i.e. the more the same people view each time). Thus a programme with an average rating of 20 and a more typical repeat-viewing percentage of only 55% would reach as many as 60% of the population in four episodes, with each viewer seeing an average of just two episodes. In the case of 'Brideshead Revisited' in Table 5.11, the people who saw the series at all saw an average of 4.4 of the 11 episodes – less than half the series.

5.7 Theoretical models

Dealing with a variety of such frequency distributions becomes a relatively complicated matter because of the number of variables that are involved such as the number of episodes, the rating levels and the repeat-viewing levels. A mathematical model to summarise such data can therefore be helpful, e.g. for prediction.

This type of knowledge will generally be of interest to those responsible for programme decisions and to students of mass communications. It also provides advertisers and agencies with important information for media planning. Given an advertising campaign of so many 'spots', the advertising agency and its client would usually like information not merely on total ratings (obtained from the regular syndicated audience measurement) but also the number of people exposed to 0, 1, 2, etc. of the advertisements. If a predictive model can be developed not only can past campaigns be evaluated, but alternative future plans can be compared.

The development of models for this general purpose is discussed in an appendix to this chapter. For our special case of regularly screened programmes we can use the Beta-Binomial Distribution – or BBD for short – to estimate the frequency distribution of viewing. It is applied here to TV programme schedules under 'stationary' conditions, i.e. when there is little or no variation in rating levels or repeat-levels.

The Fit of the BBD Model

For 'exposure distributions to different episodes of a programme under stationary conditions – when rating and repeat-viewing levels show little or no variation – the BBD model generally gives a good fit.

Table 5.13
The fit of the BBD to stationary schedules of four episodes
(observed and theoretical BBD figures)

| London Men | | **% Seeing the Programmes** | | | | |
April/May 1971		0	Once	2 times	3 times	4 times
This is Your Life	Obs.	50	15	9	13	13
	BBD	**49**	**15**	**11**	**10**	**15**
Average of 20 cases	Obs.	57	16	10	9	8
	BBD	**57**	**16**	**10**	**9**	**8**

This close fit is illustrated in Table 5.13, both for the schedule of 'This is Your Life' in the four weeks ending 9 May 1971 shown in Table 5.12 and for the average of some 20 other 'stationary' four-week schedules (covering serials, series, comedy and musical shows, quizzes, films, plays, etc.).

The model therefore serves to summarise such data succinctly. Given the average rating and repeat-viewing level, it can reproduce the full frequency distribution of exposures and, in particular, the reach and average frequency. Thus, for 'This is Your Life' the observed and predicted reach are 50 and 51 and the average frequency values are 2.48 and 2.49.

The value of such analyses of the distribution of exposures to a series of episodes is that it allows us to summarise, and hence understand, such viewing patterns better. For the typical programme in Table 5.13, viewed by just under 25% of the population in an 'average' week, we can draw conclusions such as the following:

(i)   that nearly twice as many people (i.e. 100 − 57 = 43%) would see it at least once in four weeks;

(ii) that only a third of a given episode's audience – 8% out of 25% – sees all four episodes (or only about 8/43 = 19% of those reached in the four weeks); and

(iii) that such results are normal and predictable for the general run of programmes.

The BBD model also serves to summarise many somewhat non-stationary cases to a first order of approximation, even though the theoretical basis from which the model was derived no longer applies.

Table 5.14 summarises results for more than a dozen regularly screened programmes whose ratings declined over the four weeks analysed by an average of 25%. The observed frequency distributions of exposures are still adequately summarised by the theoretical BBD (to within an average of one percentage point for the various individual schedules).

Table 5.14
The fit of the BBD for some non-stationary schedules of four episodes

London Men April/May 1971		0	**% Seeing the Programme** Once	2 times	3 times	4 times
Average of 14 non- **stationary** programmes	Obs. **BBD**	58 **57**	18 **19**	11 **11**	8 **8**	5 **5**

Furthermore, a descriptive model like the BBD, which works under a wide range of conditions, has value even in those particular cases where it does *not* fit the data, by pointing to the nature of the discrepancies. An illustration of this is provided by the analysis of the twice-weekly ITV programme 'Coronation Street' in Table 5.15.

Table 5.15
The observed and theoretical distribution for eight episodes of 'Coronation Street' in four weeks

London Men April/May 1971			**Number of Episodes Seen** 0	1	2	3	4	5	6	7	8
Coronation Street	Obs.	%	41	13	8	7	8	6	6	6	5
	BBD	%	**37**	**15**	**10**	**8**	**7**	**6**	**6**	**5**	**5**
Difference			4	−2	−2	−1	1	0	0	1	0

The average rating was 29, and the average repeat-viewing percentage 59. The theoretical BBD for such a schedule gives broadly the right picture but it differs significantly in some details. Somewhat fewer people were reached than predicted (59% versus 63%) because fewer saw only one or two episodes (21% versus 25%). These discrepancies are due to the *non-stationarity* for this series of episodes noted in Section 5.2 (Easter Monday, etc.).

Table 5.16 gives a more recent example – the fit of the BBD for the observed data for 'Brideshead' in Table 5.11. The fit is to within an average of just over one percentage point. But there is an observed five point excess of people who saw one or two episodes. This is largely due to the higher rating level of the first episode noted earlier (as can be checked directly by fitting the model to the data for episodes 2–11 only).

Table 5.16
'Brideshead Revisited': observed and theoretical distribution of episodes seen

London 1981			Number of Episodes Seen										
			0	1	2	3	4	5	6	7	8	9	10	11
Brideshead	Obs.	%	40	17	11	8	4	3	3	4	2	4	2	2
	BBD	%	43	14	9	7	5	5	4	4	3	2	2	2

5.8 Summary

Only about half – say 55% – the people who see a repetitive programme one week see the next episode in the following week. This is for popular programmes with ratings of 10 or more. For low-rating programmes the repeat-viewing levels are even lower.

There is little difference in this by type of programme or demographic group. The repeat-viewing level shows little change from the average level of about 55% even for episodes further than one week apart, implying that people watch irregularly rather than that they necessarily stop viewing a series altogether.

It follows that few people see all or nearly all the episodes in any extended series or serial. The frequency distribution of numbers of episodes seen is generally closely predictable, i.e. follows a regular pattern.

The implication is that failure to repeat-view is generally a matter of variable social habits rather than a reaction to programme content.

Appendix: the beta-binomial model

There have been a number of efforts to produce suitable models of reach and frequency. In the UK one of the earliest was that produced by the industry's previous audience measurement agency, Television Audience Measurement. In 1966 TAM produced a reach and frequency guide based on 280 schedules which had been evaluated. The guide permitted reach and average frequency to be predicted from information on the average rating of the schedule and the maximum rating.

By 1969 it was apparent that the observed reach and frequency in most schedules differed from that predicted by the 1966 TAM guide. This stimulated further work by various parties to improve the TAM predictors. In 1972 Audits of Great Britain (AGB), which had succeeded TAM in holding the industry's research contract, produced its own guide (Fawley and Fairclough, 1972). Other models were developed by Johnson and Peate (1966), Barnett and Lougher (1971) and Hulks and Thomas (1973). The last uses the Negative Binomial Distribution (NBD) to explain the distribution of frequencies.

A somewhat more direct approach is available for explaining the reach and frequency associated with regularly screened programmes. This approach estimates the frequency distribution of viewing from a Beta-Binomial Distribution, or BBD for short. The use of a BBD for media exposures was first introduced by Hyett (1958) and extended by Metheringham (1964) for print media, and has also been taken up by a number of US workers.

The use of the BBD to describe television viewing rests on two assumptions:

1. That the probability that a given person will watch that week's episode of the programme is constant from week to week (for that person) and independent of whether or not he has watched it in previous weeks.
2. That the numerical value of this probability varies from person to person (as shown by the fact that some viewers watch most episodes, and some only a few) and that this variation follows a so-called Beta-distribution in the whole population of potential viewers.

The Beta Distribution has no direct justification at this stage but it is a 'flexible' distribution which can take a variety of different shapes and hence is not a very restrictive assumption. (An earlier assumption of a similar kind in the field of consumer purchasing behaviour has recently been fully justified – see Goodhardt and Chatfield, 1973; Goodhardt *et al.*, 1984.)

The consequence of these assumptions is a mathematical formula, the BBD. With it we can calculate, from the average rating and the average repeat-percentage, how many people saw 0, 1, 2, 3, etc. episodes and, in particular, the reach and average frequency of viewing.

The first assumption implies that for a person who has a (fixed) probability p of seeing each of the episodes, the probability that he will see exactly r out of the n episodes is given by the Binomial Distribution. Thus:

$$\text{Prob } (r/p) = \frac{n!}{r!(n-r)!} \, p^r(1-p)^{n-r}, \quad O \leqslant r \leqslant n$$

The second assumption states that the probability p is distributed through the population with a probability density function given by

$$f(p) = \frac{p^{\alpha-1}(1-p)^{\beta-1}}{B(\alpha, \beta)}, \quad O \leqslant p \leqslant 1,$$

where α and β are positive constants and $B(\alpha, \beta)$ is the Beta function

$$B(\alpha, \beta) = \int_{O}^{1} x^{\alpha-1}(1 - x)^{\beta-1} dx$$

Putting these two assumptions together implies that the probability $P(r)$ that a person chosen at random from the population will see exactly r out of the n episodes is given by:

$$P(r) = \int_{0}^{1} \text{Prob}(r/p)f(p) \, dp$$

$$= \frac{n!}{r!(n-r)!B(\alpha, \beta)} \int_{0}^{1} p^{r+\alpha-1}(1-p)^{n-r+\beta-1} dp$$

$$= \frac{n!}{r!(n-r)!} \frac{B(\alpha + r, \beta + n-r)}{B(\alpha, \beta)}$$

The distribution defined by this equation is called the Beta-Binomial Distribution. It follows from elementary probability theory that in a large population the proportion of the total population who see exactly r out of n episodes will also be $P(r)$.

The BBD is defined by three parameters: the number of episodes, n, and the two constants α and β. In any practical application therefore the two parameters α and β have to be estimated. It can be shown that under the two assumptions of the BBD model given above, the proportion of the population seeing each episode (i.e. the rating of each) is equal to $\alpha/\alpha + \beta$ and the proportion of the audience of one episode who see a particular other episode (i.e. the repeat-viewing rate) is equal to $(\alpha + 1)/(\alpha + \beta + 1)$. Thus the parameters can be estimated by solving the two equations:

$$\text{average rating} = \frac{\alpha}{\alpha + \beta}$$

$$\text{average repeat} = \frac{\alpha + 1}{\alpha + \beta + 1}$$

Substituting these estimated values of α and β into the equation for $P(r)$ for different values of r gives the complete frequency distribution for the n episodes. In particular with $r = O$, $P(O)$ is the proportion seeing no episodes and so the reach is calculated as $1 - P(O)$.

The two parameters α and β of the BBD make it a very flexible distribution over the integers from O to n. For various values of the parameters it can take a wide variety of shapes from ordinary humped shapes, either symmetrical or skew, to J and reverse-J shapes and even U shapes.

6 Total TV Viewing

Just as only some viewers see all or nearly all episodes of a regular programme shown over a period of time, and many see only a few, so some people are heavy consumers of television in general and others view relatively little.

In this chapter we examine this variation in the total amount of time people spend watching television and how they divide this between channels and between programme types.

6.1 Hours viewed

In a typical week in the early eighties more than 90% of UK adults watched television and they did so for an average of about 25 hours. The figures tend to be a little lower in the summer than in the winter. There are also some demographic differences, with 'middle-class' (ABC1) adults watching somewhat less than average (about 20 hours a week) and a marked age-trend: from 15–20 hours a week for the under-35s to around 30 hours a week for the over-55s. (In late 1984 reported UK viewing figures generally rose by some 20% due to a new 'people-meter' measurement system picking up more fringe-viewing.)

Individuals viewing levels are highly variable, as illustrated in Table 6.1. About 20% watch less than 10 hours a week while at the other extreme about 30% watch more than 30 hours a week.

1984

Table 6.1
The number of hours TV viewed in a week

Adults	Hours viewed in a week							Average hours	
1980–84	0	−2	−6	−10	−20	−30	30+	per adult	per viewer
% of Population	8	2	4	7	22	25	32	**24**	**26**

70

6.2 The two major channels

Of all viewing in the UK, around 50% tends to be of ITV and just under 40% of BBC1, with the remaining 15% or so split between BBC2 (about 10%) and Channel 4 (about 5%). Each of the two major channels is seen by almost 90% of adults in the average week and by some 60%–70% in a day. But again, viewers differ a great deal in the amount they watch of a particular channel, as shown in Table 6.2.

The table shows little difference between ITV and BBC1 in terms of *light* viewers. About 40% of adults watched each channel no more than six hours a week. But ITV had many more *heavy* viewers: about 20% of adults watched the channel for 20 or more hours a week compared with only 8% for BBC1. It is this difference that produces ITV's greater share of total viewing.

Table 6.2
Hours viewed of the channel in a week

Adults 1980–84		Hours viewed in a week						Average hours per adult	per viewer
		0	−2	−6	−10	−20	20+		
ITV	%	12	8	18	15	26	21	**12**	**14**
BBC1	%	12	7	21	22	30	8	**9**	**10**

One factor associated with the large number of heavy viewers of ITV is a marked positive correlation between heavy viewing of the channel and heavy viewing of *total* television. Among heavy TV viewers (watching 35 hours or more a week) the average viewer watched 24 hours of ITV compared with 16 hours of BBC1. Relatively light TV viewers (the half of the population watching less than 20 or so hours a week) watched rather *less* ITV than BBC1 – an average of 3.8 hours ITV and 5.2 hours BBC1.

6.3 The two smaller channels

For BBC2 and Channel 4 average hours of viewing per adult are much lower than those for the two big channels: about 2.3 hours a week for BBC2 and 1.0 hours a week for Channel 4 (in 1983/84). As would be expected from these averages, very few people are heavy viewers of either channel. Table 6.3 shows

Table 6.3
Hours viewed of the channel in a week

Adults 1980–84		Hours viewed in a week					Average hours	
		0	−2	−6	−10	10+	per adult	per viewer
BBC2	%	26	32	32	8	2	**2.3**	**3.1**
Channel 4	%	51	29	17	2	1	**1.0**	**2.1**

that only 1 or 2% of adults watched 10 or more hours in a week compared with some 40% for ITV and BBC1 in Table 6.2.

The more surprising results in this table may be the percentages who had seen *any* of the channel during the week: the 'reach' of the channel. Thus, as many as three-quarters had seen BBC2 (i.e. 100% minus 26% non-viewers) and virtually half had seen Channel 4. (In 1985, as Channel 4 became more established, as many as 60–70% saw it during a week.) Although these channels have small shares of total viewing, i.e. most of their *programmes* have low ratings, viewing of the *channel* is not confined to a single minority. Instead many people watch them sometimes, for a small part of their total viewing. Even in a single day some 30% of adults might 'dip into' such a channel.

The characteristic of a small channel is not that few people watch it (although the reach figures are substantially lower than those for ITV and BBC1) but that those who do watch see rather little of it – an average of only two or three hours a week. This is a pattern that has been observed also in the US (see Chapter 7) where only channels targeted clearly at population segments (e.g. by transmitting in Spanish) deviate from the norm.

6.4 Multi-channel viewing

It is obvious from the weekly reach figures quoted above – about 90% for ITV and BBC1, 75% for BBC2 and 50% for Channel 4 – that most people watch more than one channel in a week. In fact, the average UK viewer sees three different channels in a week.

Even on a single day viewers tend to see more than one channel. In the spring of 1980 (even before the launch of Channel 4) about 80% of adults saw some television on the typical day. The average viewer saw two channels that day: 30% watched just one, 45% two and 25% all three.

Looked at in terms of those patronising a given channel that day, the pattern is striking. The vast majority also watched other channels the same day, as is shown in Table 6.4. Fewer than 20% of those who watched ITV or BBC1

Table 6.4
Viewers watching 1, 2 or 3 channels in a day

London & Lancashire 1980		Number of channels viewed		
		1	2	3
Viewers in a day of				
ITV	%	17	52	31
BBC1	%	17	53	30
BBC2	%	6	31	63

watched *only* that channel in the day. For BBC2 the figure was even lower, only 6% of those who watched the channel at all being viewers *only* of BBC2 and nearly two-thirds being viewers of all three channels. (These patterns are if anything even more striking since the advent of Channel 4 in 1983.)

Another way of looking at multi-channel viewing is in terms of hours of viewing. For the average viewer of a channel, how much of their viewing in the day is devoted to that channel and how much to *other* channels?

Table 6.5, based again on 1980 viewing patterns, shows that the average person who saw a channel at all during a single day watched around five hours of television in the day. For viewers of each of the two major channels that time was divided roughly evenly between viewing the channel and viewing other channels (60:40 in the case of ITV; 50:50 for BBC1). The average viewer of BBC2 – a slightly heavier viewer of television in total – devoted only about a quarter of his or her viewing to BBC2, about three-quarters to other channels. This illustrates dramatically the way the smaller channel is used as something to 'dip into'.

Table 6.5
Hours of viewing in the average day

London & Lancashire 1980	Any TV	Hours spent viewing The given channel	Other channels
Viewers in the average day of			
ITV	4.9	2.8	2.1
BBC1	4.7	2.4	2.3
BBC2	5.1	1.3	3.8

6.5 Viewing of programme types

We now turn from considerations of channels to examine the way people divide their total viewing between *programme types*. What emerges is that people generally view a wide range of programme types. Further, as has been seen earlier in terms of duplication of viewing for pairs of programmes, the audience does not 'segment' into groups showing marked preferences for programmes of a given type.

This can be seen by taking viewers of a given programme and examining how their *total* viewing in a week is divided between programmes of the *same* type and programmes of *other* types. Table 6.6 illustrates this analysis for a particular programme, 'Hawaii Five-O', in a given week in 1980. The average viewer of this programme spent 9% of their viewing that week watching adventure/action programmes (including 'Hawaii Five-O'), 6% watching romance, 17% watching light entertainment and so on.

Table 6.6
Viewers of Hawaii Five-O: % of their viewing time in the week spent viewing different types of programme

1980	% of time spent viewing programmes of							
	Advent. action	Romance	Light ent.	Films Plays	Sport	News & CA	Gen. int.	Misc.
Viewers of Hawaii Five-O %	9	6	17	27	13	11	9	6

not consistent

Table 6.7 examines the more general pattern. It shows how viewing hours were distributed over the programme types for the viewers of a programme of a given type. For example, viewers of an adventure/action programme devoted 6% of their viewing to programmes of that type, 6% to romance, 15% to light entertainment and so on. (In this case, viewing the original 'base' programme – like 'Hawaii Five-O' in Table 6.6 – is excluded from the analysis which therefore asks how the *rest* of the individual's viewing in the week was made up.)

The striking finding here is how little the distribution of viewing time over the programme types varies for different audiences. For the viewer of *any* given programme, about 6% of the remainder of their week's viewing was of adventure/action programmes, about 5% of romance, about 15% of light entertainment and so on. There is almost no tendency for them to focus on programmes of the same or of similar programme type.

There are variations but they are few and small. Thus several of the figures in the main diagonal of the table are very slightly above average, but only by a percentage point or two. This reflects a small tendency to view other

Table 6.7
Viewing of programme types by viewers of
programmes of the different types

London 1980		Adv. act.	Rom- ance	Light ent.	Films Plays	Sport	News & CA	Gen. int.	Misc.
		% of time spent viewing programmes of							
Viewers of the average prog- ramme of									
Adventure/action	%	6	6	15	28	15	11	10	7
Romance	%	7	7	16	26	14	12	10	7
Light entertain- ment	%	6	5	16	25	15	12	12	7
Films & plays	%	7	5	15	27	14	11	11	7
Sport	%	6	5	16	26	15	12	11	6
News & c. affairs	%	6	5	15	24	15	13	13	7
General interest	%	5	4	15	24	14	12	16	7
Average Prog- ramme*	%	6	5	15	26	15	12	13	7

Note: *Weighted by ratings

programmes of the same type. The largest effect is for viewers of general interest programmes. They spent 16% of their viewing time watching other such programmes as against an average of about 12% amongst viewers of programmes of other types.

6.6 Lighter viewers of television

Much attention over the years has centred on the behaviour of lighter viewers of television. As we have seen earlier, they do differ from heavy viewers in terms of channel choice, showing a marginal preference for BBC1 over ITV. Do they differ in other respects?

A common view of television audiences is that people who do not view much must be selective in what they watch. The word 'selective' here would usually be taken to imply that these people watch minority-interest programmes, especially, perhaps, ones with a cultural or specialist appeal, which therefore only attract small audiences, rather than the popular (and high-rating) type of entertainment programmes.

In practice this does not occur. Lighter viewers of television – watching up to

20 hours a week – devote 23% of their viewing to low-rating programmes (with ratings less than 10). This is much the same as is found among all other viewers, 21% of whose viewing is of such low-rating programmes.

What is special about lighter viewers is that little of their viewing (about 20%) takes place before 6 pm. Among those watching over 40 hours a week, in contrast, 36% of viewing is before 6 pm. So, perhaps the main characteristic of light viewers is their tendency to watch mainly at peak-time. That is why it is peak-time.

Another feature of 'selectiveness' might be that light viewers are more *regular* viewers of the programmes they do watch week by week. They may not watch much but perhaps they know what they like.

However, the actual results again go in the contrary direction. In Chapter 5 we saw that a little over 50% of the audience of a typical programme would see it again in the following week. But for light viewers, the repeat-viewing percentage is lower, not higher, at about 40% only.

6.7 Summary

Viewers differ greatly in the amount of television they watch. Something like one-in-three watch more than 30 hours a week (or four hours a day) and, at the other extreme, one-in-five watch only up to 10 hours a week (an hour or so on the average day).

Similarly, amounts of viewing vary widely for the individual channels (although very few people are heavy viewers of the small channels: BBC2 and Channel 4). Most people, however, see *some* of each channel's output and multi-channel viewing is the norm. Even in a single *day* most viewers watch two or more different channels.

Viewers also tend to spread their viewing over a wide range of programme types. There is little or no special tendency for those who watch any given programme to devote the remainder of their viewing to other programmes of the same or similar type.

Lighter viewers of television watch slightly more BBC than ITV, in contrast with the marked preference for ITV among heavy viewers. But light viewers do not seem to be especially 'selective' in their viewing. They tend to watch mostly at peak-time when they show no special tendency to watch specialist (i.e. low-rating) programmes. They also show no sign of being more regular viewers of the programmes that they watch. They do not appear to be selective in either sense but just to watch less.

7 Audience Flow in the US

In this chapter we describe some comparable results in the United States. A good deal of work on US ratings data has been done in recent years which has confirmed and extended the initial findings in the first edition of this book. Here therefore, we set out our initial findings which still illustrate that the approach of the preceding chapters, developed for the UK, can also be applied to the study of television audiences elsewhere. Less work is now needed to establish whether or not the same results recur.

Despite differences in the US and UK television scenes, the finding is that the main patterns of audience flow checked so far are the same as in the UK. For example, the duplication of viewing law between different programmes operated both in the sixties in relatively 'small' regions such as Birmingham, Alabama and Las Vegas which then had only two or three main channels (e.g. Ehrenberg 1966) and in 1974 in the New York City area where up to 10 or so channels could be received.

Additional results are that repeat-viewing between different episodes within a week was then about 55% for the three' national networks, roughly the UK level. More recently, repeat-viewing is lower still, below 50% (Barwise 1986, Barwise *et al.*, 1982, Ehrenberg and Wakshlag 1986). Also as in the UK, the 'inheritance effect' operates for consecutive programmes on the same day.

We illustrate the detailed findings for the US, as in our first edition, with results from a sample of 1779 female heads of household in the New York City area in January and early February 1974. (The sampling was spread over four weeks, but only one week's viewing is measured per informant.)

The results relate to the six most popular stations in the New York City area. These are the affiliates of the three national networks (CBS, NBC, and ABC), plus three stations (WOR, WNEW and WPIX) which either belong to regional networks or are independent. Ratings for the educational station WNET were too low to be usefully included in the analysis.

7.1 Repeat-viewing of different episodes in the same week

We start with an analysis of the repeat-viewing of different episodes of the same programme. The situation analysed here is that of strip-programming, i.e.

77

different episodes of the same programme being shown on successive week-days. This occurs in two different ways. Firstly, repetitive programmes are shown from 4 to 7.30 pm on the three national networks, if we also count local and general news in this category. Secondly, the other three stations show reruns of old programmes and old films in the afternoon and, in some cases, throughout the evening as well, e.g. repeats of 'The Lucy Show' originally shown on NBC 20 years previously.

On the national networks repeat-viewing for different week-days averages at roughly 55% – the same level as found in the UK for week-by-week repeat-viewing. Table 7.1 shows specifically the extent to which episodes on a Friday were watched by viewers on other week-days. The overall average is 53%. Thus only about half the people who see one episode of a programme tend to see another. The table also shows that there is no substantial erosion of this repeat-viewing level for days further apart. There is a suggestion in Table 7.1 that

Table 7.1
Repeat-viewing within the week: NETWORKS
(% of Monday to Thursday audiences of a programme who watched the Friday episode of that programme)

New York Housewives Jan–Feb 74		% Viewing on Friday of the Audience on				
		Mon.	Tue.	Wed.	Thu.	Av.
WCBS	4.00 Secret Storm	72	65	69	69	**69**
	4.30 Mike Douglas	65	64	61	71	**65**
	6.00 Ch 2 News-6	64	67	60	72	**66**
	7.00 CBS Eve. News	62	63	63	65	**63**
	Average	**66**	**65**	**63**	**69**	**66**
WNBC	4.00 Somerset	71	76	66	68	**70**
	4.30 Movie Four	15	20	18	29	**21**
	6.00 Sixth Hour	51	50	51	58	**52**
	7.00 NBC Night News	45	50	45	49	**47**
	Average	**45**	**59**	**45**	**51**	**47**
WABC	4.00 Love Am. Style-D	38	47	39	56	**45**
	4.30 4.30 Movie	29	26	43	51	**37**
	6.00 Eywtns News 6	59	55	56	58	**57**
	7.00 ABC Eve. News	52	50	43	49	**48**
	Average	**45**	**44**	**45**	**53**	**47**
Overall Average		**52**	**53**	**51**	**58**	**53**

repeat-viewing for consecutive days – Thursday and Friday – is a few percentage points higher. (The Arbitron measurement week runs from Wednesday to Tuesday so that the Monday and Tuesday results in this table refer to the extent to which the viewers then had seen the previous Friday's episode.)

There appears to be some fairly substantial variation about the overall average of 53% in the repeat-viewing levels for different programmes, and possibly for different networks. Repeat-viewing for the four programmes on WCBS (the New York CBS affiliate) are all in the 60s whereas the levels for the NBC and ABC afternoon films are rather low. But it has not been shown whether these variations are generalisable.

Repeat-viewing levels for the other New York stations studied – WNEW, WOR, WPIX – are mostly lower than for the networks. This is illustrated in Table 7.2 for WNEW where strip-programming continues right through the evening. From 4 to 7 pm the repeat levels average at 35%, compared with 53% in Table 7.1 for the major networks. Later in the evening the repeat levels are sometimes lower still, possibly due to the competition of the high-rating network programmes then.

Table 7.2
Repeat-viewing within the week: WNEW
(% of Monday, Tuesday and Thursday audiences of a programme who watched the Friday episode of that programme)

New York Housewives Jan–Feb 74		Mon.	Tue.	Thu.*	Av.
		% Viewing on Friday of the Audience on			
WNEW	4.00 Bugs Bunny	33	20	16	**23**
	4.30 Lost in Space	31	33	42	**35**
	5.30 Flintstones	30	25	40	**32**
	6.00 Lucy Show	48	38	46	**44**
	6.30 Bewitched	39	38	39	**39**
	7.00 Mission Imposs.	35	42	40	**39**
	8.00 Dealers Choice	9	15	19	**14**
	8.30 Merv Griffin	30	33	34	**32**
	10.00 10 O'Clock News	43	40	40	**41**
	11.00 Step Beyond	24	29	26	**26**
	11.30 11.30 Movie	16	21	13	**17**
	Average	**31**	**30**	**32**	**31**

Note: *Wednesday not tabulated

As background to this analysis, Table 7.3 sets out the WNEW ratings over the week-days. The ratings are mostly steady from day to day. But this is only on the surface – it does not mean that the audiences consist of the same people every day. As we have seen from Table 7.2, only about a third of those viewing a particular programme are repeat-viewers from one day to another. (The ratings are similarly steady for the networks – as was illustrated in Table 2.2 – but still only about half the viewers are the same from day to day.)

Most of the ratings for WNEW are very low. This may help to explain the low repeat-viewing levels in Table 7.2. As noted in Chapter 5, repeat-viewing tends to decrease with rating in the UK.

Table 7.3
Week-day ratings for WNEW programmes

New York Housewives Jan–Feb 74		% HW's Viewing				
		Mon.	Tue.	Thu.*	Fri.	Av.
WNEW	4.00 Bugs Bunny	1	1	1	1	**1**
	4.30 Lost in Space	1	1	1	1	**1**
	5.30 Flintstones	1	2	1	1	**1**
	6.00 Lucy Show	2	3	2	2	**2**
	6.30 Bewitched	3	3	3	3	**3**
	7.00 Mission Imposs.	5	5	5	4	**5**
	8.00 Dealers Choice	3	1	2	1	**2**
	8.30 Merv Griffin	6	6	5	5	**6**
	10.00 10 O'Clock News	9	10	10	8	**9**
	11.00 Step Beyond	2	2	2	2	**2**
	11.30 11.30 Movie	1	1	1	2	**1**
	Average	**3**	**3**	**3**	**3**	**3**

Note: *Wednesday not tabulated

Another possible influence on repeat-viewing levels was also noted in Chapter 5. Repeat-viewing analyses between two episodes are simple when the two ratings are steady but problems arise when they are not. For example, for two episodes with ratings 4 and 3, only 75% of the first audience could possibly see the second episode. And with low-rating programmes as for WNEW here, small differences in rating levels become proportionately large. Thus two ratings of 2 could reflect actual audience levels as different as 2.5 and 1.5, allowing a maximum possible repeat-viewing percentage of only 60%. With sample data, rating differences as such can arise simply because of sampling errors.

7.2 Duplication between different programmes

The level of audience overlap between different programmes in these New York data is usually lower than the level of repeat-viewing for different episodes of the same programme, as found in the UK also. Instead of averaging at 30–50%, audience duplication for different programmes on different days is generally below 20%. The duplication is still directly proportional to the programmes' ratings, as in the UK, and the duplication of viewing law of Chapter 2 again applies.

For the three national networks in the 1974 New York data, the duplication coefficient – the ratio of the duplicated audience to the rating – is mainly of the order of 1.5 or 1.6 for programmes on the same channel and about 1.1 or 1.2 for programmes on different channels. Thus the degree of channel loyalty (the within-channel factor of 1.5 or 1.6) is marginally smaller than that in the UK recently (1.7 or so). But the amount of switching between channels is relatively higher – 1.2 versus 0.9 in the UK. We now describe the results in more detail.

Table 7.4 gives the detailed audience duplication figures for ABC programmes on Thursdays and Fridays – a typical example for a US network. The general tendency is for the Friday ABC programmes to be more popular among Thursday ABC viewers than they are in the population as a whole. Virtually all the Friday duplications in the table are higher than the Friday ratings, usually by a factor of 1.5 to 1.7.

Table 7.4
Duplication of viewing between programmes on the same channel
(% of viewers of one programme who also watch another on another day)

New York Housewives Jan–Feb 74		Who also viewed ABC on Friday at				
		7.30	8.00	9.00	11.00	11.30
Viewers of ABC on THURSDAY at						
4.00 Love Am. Style-D	100%	17	13	23	23	10
4.30 4.30 Movie	100%	28	12	18	14	4
6.00 Eywtns News 6	100%	23	11	16	22	7
7.00 ABC Eve. News	100%	31	12	19	23	7
7.30 Animal World	100%	20	7	12	15	3
8.00 Chopper One	100%	14	8	26	21	7
9.00 Kung Fu	100%	14	12	21	20	7
10.00 Strts Sn Frn	100%	13	8	24	24	6
11.00 Eywtns News 11	100%	16	9	21	(43)	(13)
11.30 Wide World Entert.	100%	16	11	18	(36)	(17)
Average		19	10	20	20*	6*
1.6 × Rating		**19**	**10**	**21**	**18**	**6**
Rating		12	6	13	11	4

Note: *Excluding late night cluster

For example, the 8 pm Friday programme (which was usually 'Brady's Bunch' in the four weeks analysed) was seen by roughly 10% of the viewers of any ABC Thursday programmes, compared with its rating of 6: a duplication factor of just over 1.7. The 9 pm Friday programme ('the $6 Million Man' and others) was seen by about 20% of viewers of any Thursday programme compared with its rating of 13 – a duplication factor of 1.5, and so on.

The individual figures in each column of the table vary somewhat but most of the differences from the column average are relatively small (on average about 3 percentage points) and they appear to be largely irregular. However, there are two exceptions, both having higher duplication.

The first exception is for the Friday 7.30 programme. It has relatively high duplications with the 7.30 pm or earlier programmes on the preceding day. This is a common finding for all week-days on the three networks. It is probably due to an 'inheritance effect' from the 7 pm news when the strip-programming type of high day-by-day repeat-viewing levels operate, as discussed earlier.

The second exception is the late-night cluster of high duplications, seen in the bottom right-hand corner of the table. Duplication between the audiences at 11 pm or later on each of the two days is markedly higher than shown by the rest of the table. This is again a general phenomenon on all networks in the US and one we have also seen for the UK (Chapter 4). The cluster is thought to be due to many people habitually going to bed between 10 and 11 so that those who stay up tend to be the same people night after night. As argued in Chapter 4, the late-night ratings are 'too low' and hence make the duplication appear high. If late-night viewers were expressed as a percentage of those then available to view, the 'ratings' would be higher. The late-night duplications would then no longer appear so high but more in line with the patterns observed for the rest of the evening.

Direct numerical evidence of people's availability is lacking but the duplication of late-night viewers on one day for programmes at 10 pm or earlier on the next day does show that, as in the UK, there is nothing abnormal about the late-night viewers as such. They are not especially heavy viewers of television in general. If they were their duplication levels would be high between different programmes generally. But it is only late in the evening that this happens.

The general tendency of audience duplication for the three main networks therefore is for viewers of one programme to be about 60% more likely to watch another programme on the same channel on another day than is the public in general, i.e. the percentage of duplicated viewers is about 1.6 times the rating. The duplication coefficients for the three network channels in the 1974 New York data are: WCBS 1.48, WNBC 1.58, WABC 1.64. The striking feature is the similarity rather than the differences of the three values, all about 1.6.

The situation illustrated in Table 7.4 was for two consecutive days. There is little change in the duplication coefficients for pairs of days further apart – at most a slight decrease. The average for the three networks is

consecutive days – 1.60
1 day apart   – 1.58
2 days apart   – 1.55
3 days apart   – 1.54

The trend is consistent but numerically small.

Comparable results about within-channel duplication for the three independent New York stations analysed here hardly occur because of the prevalence
of strip-programming. Instead of duplication between different programmes at
the same time on two days the situation is dominated by repeat-viewing of
different episodes of the same programme. Repeat-viewing being generally
much higher than normal duplication levels, the audience overlap for two
programmes shown at different times is often still affected by the high repeat
levels operating for programmes at the same times, through the inheritance
effect (see below). Furthermore, individual results tend to be highly variable
because of small rating levels for the independent stations and, hence, small
subsamples in the data.

Duplication between channels

Audience overlap between programmes shown on different days on different
channels tends to be fractionally higher than the ratings, by a factor of 1.1 or
perhaps 1.2. Thus, viewers of an ABC programme are, if anything, very
slightly more likely to see a CBS programme the next day than is the population
as a whole.

Table 7.5 illustrates these cross-channel results, showing the percentages of
viewers of ABC programmes on a Wednesday who saw the CBS programmes
on the following Tuesday. In this particular table the duplications nearly equal
the ratings to within a few percentage points. But in other cases the
duplications tend to be about 10 or 20% higher, giving the duplication
coefficient of 1.1 or 1.2 mentioned earlier.

These findings – within-channel duplication coefficients of about 1.5 to 1.6
but between channels of about 1.1 – imply that channel loyalty exists in the
main networks but to a lesser degree than in the UK. This may be due to the
greater similarity of the US network offerings or perhaps to the stations'
tendency to schedule programme-changes at the same times (unlike ITV and
BBC1 in the UK). (Between-channel duplications involving the smaller
stations have not yet been effectively summarised because of statistical
problems arising from the small samples of viewers that are involved.)

Table 7.5
Duplication of viewing BETWEEN channels

New York Housewives Jan–Feb 74		*Who also viewed CBS on TUESDAY at*				
		7.30	8.00	9.30	11.00	11.30
Viewers of ABC on WEDNESDAY at						
4.00 Love Am. Style-D	100%	16	34	10	7	4
4.30 4.30 Movie	100%	14	22	11	2	3
6.00 Eywtns News 6	100%	11	26	13	3	3
7.00 ABC Eve. News	100%	12	29	14	6	4
7.30 Strange Places	100%	12	29	10	5	2
8.00 Wed Mv of Wk	100%	8	24	12	7	3
10.00 Doc Elliot (etc.)	100%	13	28	12	6	5
11.00 Eywtns News 11	100%	10	24	15	5	3
11.30 Wide World Entert.	100%	9	23	14	7	2
Average	100%	12	27	12	5	3
Rating	100%	14	26	11	7	3

7.3 The inheritance effect

It is well known that for pairs of programmes shown on the same channel on the same day there is an 'inheritance' effect, or 'lead in' as it is called in the US. The audience duplication is larger because people stay tuned to the same channel when a programme ends, or tune in to the channel early to be sure of seeing a favourite programme there later (i.e. a 'lead out' effect).

Table 7.6 shows typical same-day audience duplication figures for pairs of CBS Friday programmes. Except for some of the late-afternoon programmes and the 6 and 7 pm news shows, all but one of the overlap figures are less than 50%. Thus, it is not a case of most people 'being too lazy' to switch channels (or to switch off) once the programme they are viewing has finished.

However, the duplications are nearly all higher than the ratings shown at the bottom of the table. Thus viewers of one CBS programme are more likely than non-viewers to watch any other CBS programmes that day. But in most cases this is no higher than the normal positive audience duplication between two programmes on the same channel on different days. Table 7.7 allows for this general 'channel loyalty' effect by subtracting the predicted between-day duplications.

As in the UK (Chapter 5), the results show three main features:

Table 7.6
Same-day duplications: CBS on Friday

New York Housewives Jan–Feb 74		Who also Viewed CBS at								
		4.00	4.30	6.00	7.00	7.30	8.00	8.30	11.00	11.30
Viewers of CBS at										
4.00 Secret Storm	100%	100	**60**	51	38	13	14	13	9	8
4.30 Mike Douglas	100%	**22**	100	**55**	35	15	15	14	10	4
6.00 Ch 2 Nws-6	100%	16	**48**	100	**61**	24	22	14	15	5
7.00 CBS Eve. Nws	100%	13	32	**64**	100	**34**	18	15	15	5
7.30 Secrets Deep	100%	6	19	36	**48**	100	**33**	22	12	8
8.00 Dirty Sally	100%	5	16	26	21	**27**	100	**40**	15	7
8.30 CBS Fr Nt Mv	100%	4	12	14	14	15	**33**	100	**24**	8
11.00 Ch 2 Nws 11	100%	5	14	24	22	13	20	**39**	100	**28**
11.30 CBS Lt Movie	100%	8	12	17	16	17	19	27	**56**	100
Rating		4	12	14	13	9	11	14	9	4

Table 7.7
Lead-in effects
(Differences between observed same-day duplications and 'between-day' estimates)

New York Housewives Jan–Feb 74	CBS Friday									
	4.00	4.30	6.00	7.00	7.30	8.00	8.30	11.00	11.30	**Av.**
CBS Friday										
4.00 Secret Storm	–	**41**	29	17	−2	−4	−9	−5	1	**8**
4.30 Mike Douglas	15	–	33	14	0	−3	'−8	−4	−3	**5**
6.00 Ch 2 Nws-6	9	**29**	–	40	9	4	−8	1	−2	**10**
7.00 CBS Eve. Nws	6	13	**42**	–	19	0	−7	1	−2	**9**
7.30 Secrets Deep	−1	0	14	**27**	–	15	0	−2	1	**7**
8.00 Dirty Sally	−2	−3	4	0	**12**	–	18	1	0	**4**
8.30 CBS Fr Nt Mv	−3	−7	−8	−7	0	**15**	–	10	1	**0**
11.00 Ch 2 Nws 11	−2	−5	2	1	−2	2	**17**	–	21	**4**
11.30 CBS Lt Movie	1	−7	−5	−5	2	−3	5	**42**	–	**4**
Av. of adjacent programmes	15	35	37	33	15	15	18	26	21	24
Av. of adj.-but-one prog.	9	13	22	7	4	1	3	1	1	7
Av. all others	0	−4	−2	2	0	−2	−8	−2	−1	−2

– the inheritance effect exists primarily between adjacent programmes;
– for programme pairs separated by more than one other programme it is generally negligible, if it exists at all;
– the effect varies in size (perhaps due to different programmes or times of day).

More work is needed to establish what factors determine the actual size of the inheritance effects and to differentiate between 'lead out' and 'lead in'.

7.4 Summary

The results shown in this chapter indicate that many of the features of audience flow in the UK also occur in the US. The larger number of channels and the structuring into three main national networks, regional networks and a variety of independent stations do not in themselves seem to lead to radically different viewing patterns. Repeat-viewing, audience duplication, channel loyalty and inheritance effects are largely as in the UK. There is relatively more switching among the three main networks, perhaps because their programme and scheduling policies are less distinct than those of the four channels in the UK.

The main difference from the UK is due to a particular aspect of programming policy, namely the high incidence of strip-programming within the week. This means that a particular channel will often have a much higher audience duplication from one day to the next because this is governed by the nature of repeat-viewing. Two episodes of a programme will have 30–60% of their audience in common, whereas duplication of viewing between different programmes is seldom greater than 20% in the US. (The duplication of viewing law still operates, but because of the larger number of channels in the US, rating levels for any one programme tend to be smaller than in the UK. Hence, duplication tends to be numerically smaller as well, even though the duplication coefficients of the order of 1.6 or so for the main networks are roughly similar to the UK figures.)

The studies so far carried out in the US have shown that despite differences in the TV scene, the approach to the analysis of audience flow developed in the UK is applicable to the US and tends to produce the same kind of simple and generalisable results. The implication is that the approach is also worth following in other countries (including ones where only one channel is available). Indeed, even if audience flow patterns in some other countries should turn out to be very different, the present approach of examining repeat-viewing and audience duplication levels would help to pinpoint both the existence and the nature of such differences quickly and efficiently.

8 Audience Appreciation

So far in this book we have been concerned with people's viewing behaviour. This is a natural place to begin research into television. We need to know what people do before we can seek to explain their reasons for doing so or the effects of their actions. But there is also widespread demand for measures that go beyond counting audience size: for what are sometimes called 'qualitative ratings'.

Both the BBC and ITV interests in the UK have therefore collected both regular and *ad hoc* information on audience reactions. Much of the work refers to particular programmes at particular times and yields few results of general significance. More general findings have, however, emerged from analysis of data from the Audience Reaction Assessment system (AURA). This system was initiated by the IBA and is now run – in parallel with audience measurement – by BARB. Most of the work summarised below has taken place since the publication of our first edition, at which time our comments were mainly tentative.

8.1 The demand for measures of appreciation

Information about audience size and patterns of viewing is perhaps not enough if we are to learn why people view particular programmes and how much pleasure or value they obtain from them. Thus the level of repeat-viewing which a regular programme attracts need not be a complete guide to satisfaction. Repeat-viewing could be affected by other factors like the opposing programmes, family influence and the pull of the individual's other activities. What is more, the differences in the repeat-viewing levels of different programmes are generally not very large and are mainly related to rating level, as we have seen in Chapter 5. They therefore provide no effective guide to viewers' satisfaction in the aggregate.

Ratings vary a great deal more than do repeat-viewing levels but are obviously more a measure of mass appeal (i.e. other viewers' interest in the programme) than of the individual viewer's satisfaction. They can also be greatly influenced by extraneous factors like time of day, competitive programming on the other channels, inheritance effects and channel.

More direct indicators of viewers' appreciation of different programmes are therefore desirable. They should also shed light on more detailed aspects of audience reactions – a producer may wish to know which aspects of his programme people appreciated, how far the points made had been understood and whether the programme achieved the objectives intended.

Programme producers and planning executives are not the only ones who require additional information to ratings. Critics of television have often attacked or defended it on grounds which have little to do with the viewing figures as such. The Pilkington Committee (1960), for example, had little time for ratings:

> It is by no means obvious that a vast audience watching television all the evening will derive a greater sense of enjoyment from it than will several small audiences each of which watches for part of the evening only. For the first may barely tolerate what it sees: while the second might enjoy it intensely.

The Committee was interested in the 'quality' of the programmes, their social and psychological effects and other aspects.

Even manufacturers and their advertising agencies, who generally receive most of the blame for the dominating influence of the ratings, require data to supplement ratings figures. They want to know about the attention paid to programmes, viewers' understanding, perception and recall of commercials, and how the advertising affects attitudes and the formation of intentions to buy.

8.2 The Appreciation Index

The AURA system uses a comparatively simple measure of audience satisfaction, the Appreciation Index or AI. Viewers are asked to report the programmes they see in a week and to indicate their appreciation of each of them by marking it in one of six categories. These run from 'Not at all interesting and/or enjoyable' (scored 0) and increase in steps of 20 to 'Extremely interesting and/or enjoyable (scored 100). By averaging the scores of the different panel members, an average Appreciation Index (AI) running from 0 to 100 is obtained for each programme seen. The advantage of this simplicity is that a viewer can be asked to comment on a large number of programmes of varying types.

Viewers score only programmes which they have viewed that week and are unlikely to feel (or to say that they feel) that all of these have been a complete waste of time or positively unpleasant. As a result, average values mostly lie in the upper half of the range. Table 8.1 gives some typical examples of the distribution of individual scores and average AI values.

Table 8.1
Some percentage distributions of individual appreciation scores and corresponding average AI values
(% of viewers giving scores)

London Adults 1972		0	20	*Individual AI Scores* 40	60	80	100	*Average* AI
Viewers of								
Pot Black	%	0	0	0	2	51	47	**88**
Colditz	%	1	0	0	16	41	41	**84**
Crossroads	%	0	1	3	26	34	36	**80**
News at Ten	%	0	0	3	27	43	26	**78**
Wrestling	%	8	0	1	25	33	34	**75**
Mr. Trimble	%	7	0	0	23	51	19	**73**
Coronation Street	%	3	6	6	23	34	24	**71**
Top of the Pops	%	8	4	8	41	27	11	**61**
Candid Camera	%	13	10	13	35	16	12	**53**
Play for Today	%	13	25	6	28	10	10	**48**

These examples show that the Index is able to discriminate between more and less popular programmes, at least to some extent (i.e. a range from 48 to 88). The results that follow will show that this ability arises from discrimination by individual viewers and not just because some programmes attract viewers who are generally more likely to indicate high appreciation. We will also report experimental results showing that the apparently naive approach of asking people to score a programme in terms of its being 'interesting and/or enjoyable' copes rather well with the problem that different types of programme will produce different types of satisfaction.

8.3 Appreciation Index and audience size

In general there is no correlation between the AI and audience size (rating level). Table 8.2 illustrates this for a set of 11 programmes which featured in a particular analysis. Ratings (here the percentage of the AI panel who recorded having viewed the programme) vary from 37% down to 12% but the AI scores show no parallel trend. In these cases they were generally close to the average of 76, with just two somewhat higher scores at 84.

The lack of simple correlation here need not be surprising. The AI scores are given by those people who viewed the programme in question. There is no particular reason why those who watched a programme seen by few other people should like it less than those who saw a high-rating programme. On the other hand, it could have been that at least some low-rating programmes are seen by small audiences who enjoy them far more intensely than the general run of programmes (as suggested by the Pilkington Committee). Alternatively they might be seen by heavy viewers who do not actually like them very much but cannot switch off. We now know that neither of these extremes appear to be common.

Table 8.2
Audience size and AI score

London Adults 1972	**Rating**	**AI Score**
Man at the Top	37	84
Budgie	32	75
David Nixon Show	34	76
Smith Family	31	77
Star Trek	24	72
Saturday Variety	30	72
Parkinson	31	84
Man Outside	25	74
World in Action	19	76
This Week	18	77
Panorama	12	77
Average	27	76

More detailed analysis shows, however, that this overall pattern hides two small but systematic trends. These operate in opposing directions to produce the overall picture of no correlation. First, there is an association between AI and programme type where more serious programmes, with lower ratings on average, gain higher AI scores. This pattern is illustrated in Table 8.3 for the eight main programme types identified in AURA reports.

The three 'serious' programme types in the top half of the table have relatively small audiences but higher AI averages. In contrast, the four more 'popular' programme types in the bottom half of the table have higher ratings and lower AIs. (The relatively low rating for sport may be largely induced by scheduling in these particular figures; nonetheless, it goes with high average AI.)

To an extent, this pattern must reflect the fact that the overall 'interesting and/or enjoyable' scale is applied differently to serious and light programmes (see below). But a more important explanation is probably that to watch a

Table 8.3
*Average AI and average adult ratings for different
programme types*
(Lancs Wk27, London Wk28, Anglia Wk29 – 1977)

Adults (16+)	Av. AI	Av. Rating
Sport	81	11
Gen. interest/information	76	7
Regular news/curr. affairs	76	10
Weekend news	76	9
Romance	72	25
Comedy/light entertainment	70	19
Adventure/action	70	18
Films/plays	70	15
Average	74	14

more demanding serious programme for long enough to feel able to judge it calls for a greater 'pay-off' in terms of satisfaction than is required when watching an undemanding programme.

The second, and opposing, trend emerges when we look at individual programmes within a type. Here we find positive correlation: the programmes that are watched by more people provide greater satisfaction to their viewers. In both senses they are more 'popular'.

The pattern is most clear for higher-rating programmes. For peak-time programmes on ITV and BBC1 the correlation between ratings and AI scores averages about 0.6 once we have allowed for the general tendency for more serious programmes to achieve higher AIs. The relationship has been tentatively summarised by two equations:

$$\text{information programmes:} \quad AI = 70 + 0.33 \text{ (rating)}$$
$$\text{entertainment programmes:} \quad AI = 60 + 0.33 \text{ (rating)}$$

Both equations have a slope of 0.33. This means that if we consider different programmes within each major type, as the rating goes up by 10 points the AI goes up by 3 or 4 points on average. But information programmes with a given rating tend on average to have an AI some 10 points higher than that for entertainment programmes with the same rating.

Results for individual programmes show a great deal of scatter around the underlying relationships, as evidenced by the 0.6 correlation between AI and rating within programme type. Some of this scatter is due to sampling variability. But some systematic patterns seem to warrant more study. For example, it seems that different episodes of the same programme – seen and rated by different people – deviate in a consistent way. Some programmes of a given type are liked more than others by their viewers, even after allowing for audience size.

For lower-rating (e.g. off-peak) programmes the correlation between ratings and AI scores within programme type is still positive. But it is even lower, averaging only about 0.2. Here we would expect the small samples involved to have more impact but there could well be explanations in terms of programme characteristics to be discovered.

8.4 Appreciation Index and repeat-viewing

We might expect that people who enjoyed a programme would be especially inclined to watch the next episode of that programme. Hence we might look for possible correlation between the AI score and the level of repeat-viewing. Table 8.4 takes the same 11 programmes as in Table 8.2 and compares their AI scores with their repeat-viewing levels (i.e. the percentage of viewers in the AI panel in one week who watched the programme again two weeks later).

Table 8.4
Repeat-viewing and AI scores

London Adults 1972	*Average Repeat*	*Average AI Score*
	%	
Man at the Top	76	84
Budgie	72	75
David Nixon Show	69	76
Smith Family	68	77
Star Trek	61	72
Saturday Variety	60	72
Parkinson	59	84
Man Outside	58	74
World in Action	50	76
This Week	46	77
Panorama	43	77
Average	60	76

The repeat-viewing level decreases systematically in the table. This clearly parallels the trend in the rating levels of these 11 programmes shown in Table 8.2 and is in line with the correlation between ratings and repeat-viewing in the JICTAR data noted in Chapter 5. But since the AI scores for the programmes in Table 8.4 hardly vary, there is no correlation between AI score and repeat-viewing here. This may seem surprising. More recent analyses of programmes with a greater range of AI scores have suggested some association between repeat-viewing level and AI. But the gradient in AI levels is still quite small.

It seems that the Appreciation Index does not measure the same kind of thing as the incidence of repeat-viewers, or hardly. Part of the explanation is probably that we are dealing with repetitive programmes. Someone who really does not greatly care for a particular programme will probably not even have seen the first of a pair of episodes that we are analysing, let alone the second.

8.5 Appreciation Index and individual repeat-viewing

A more positive relationship arises when we dissect the audience of any single programme in terms of their individual appreciation and repeat-viewing. Taking 'Man at the Top', the first of the 11 programmes just considered, we can break its audience down into the different AI scores the viewers gave to the programme (grouping those scoring 0, 20, 40 and 60 together because of the small numbers of viewers giving these scores). We then find that there is a corresponding gradient in the incidence of repeat-viewing:

AI score	0–60	80	100
% repeat-viewers	62	72	85

Thus, of those who gave a low AI score to this programme, only 62% saw it again next time (i.e. two weeks later in the particular bi-weekly situation in which AI measurements are made), whereas of those who appreciated the programme highly (an AI score of 100), 85% saw it again.

This pattern occurs not only for this programme, which had a rather high repeat-viewing level, but also for virtually all the other 10 programmes in the analysis, as is shown in Table 8.5. (The only exceptions are the last two programmes which had the lowest ratings, 18 and 12, so that sample numbers for the analysis in Table 8.5 are small and unreliable.)

These results indicate that the higher the AI score a person gives to a programme, the more likely he is to watch another episode of that programme. However, this is not a simple causal link. Someone giving a high AI score to a programme is also more likely to have seen *preceding* episodes of the programme.

This is shown in summary form in Table 8.6 which compares the average repeat-viewing percentages from the bottom row of Table 8.5 with the corresponding average percentage of viewers of a programme who had seen the episode two weeks before. The correlation is therefore between AI score and frequency of viewing.

Table 8.5
Repeat-viewing by AI score
(% of viewers giving the stated AI score who are
repeat-viewers)

London Adults 1972	AI Score		
	0–60	80	100
	%	%	%
Repeat-viewers of			
Man at the Top	62	72	85
Budgie	48	76	88
David Nixon	57	71	82
Smith Family	57	66	83
Star Trek	55	66	65
Saturday Variety	48	66	68
Parkinson	53	59	66
Man Outside	53	57	62
World in Action	44	54	57
This Week	48	42	48
Panorama	25	55	35
Average repeat-viewers	50	62	67

Table 8.6
Seeing the preceding and succeeding episodes by AI score
(% of viewers giving the AI score who also saw the preceding and succeeding
episode)

Average of 11 Programmes 1972	AI Score Given to Current Episode		
	0–60	80	100
	%	%	%
Seeing previous episode	53	65	73
Seeing succeeding episode	50	62	67

Repeat, new and lapsed viewers

This correlation between AI scores and frequency of viewing can also be
demonstrated by looking at the scores given to a programme by three different
categories of viewers. Thus for any two episodes of a programme we have:

'repeat-viewers', i.e. those who saw both episodes;

'new' viewers, i.e. those who saw the second but not the first (although they may have seen yet earlier episodes);

'lapsed' viewers, i.e. those who saw the first but not the second episode.

Repeat-viewers of 'Man at the Top' gave an average AI score of 87 but 'new' and 'lapsed' viewers liked the programme somewhat less, giving AI scores averaging in the mid-70s. This pattern holds for most of the 11 programmes, as shown in Table 8.7.

Table 8.7
Audience Appreciation Index by 'repeat', 'new' and 'lapsed' viewers

London Adults 1972	Type of Viewer		
	'Repeat'	'New'	'Lapsed'
AI Score for			
Man at the Top	87	78	75
Budgie	80	61	60
David Nixon	79	67	70
Smith Family	81	69	69
Star Trek	75	73	62
Saturday Variety	75	67	71
Parkinson	86	84	83
Man Outside	76	69	74
World in Action	77	75	71
This Week	77	73	76
Panorama	81	71	75
Average	79	72	71

The similarity of the average AI scores given by 'new' and 'lapsed' viewers confirms the results in Table 8.6 that a viewer's expressed appreciation of a particular episode of a programme does not seem to be more associated with his future viewing than with his past.

The general conclusion would therefore seem to be the rather unexciting one that the more enjoyable a person finds a programme the more often he is likely to watch it. However, the situation may not be quite that straightforward. Firstly, the effect may frequently be the other way round – the more often someone watches a programme, the more enjoyable he finds it. Secondly, it appears that heavy viewers of television generally tend to give somewhat higher AI scores than do light viewers. But repeat-viewers of a particular programme usually include an above-normal proportion of heavy TV viewers.

So the heavy viewer's higher appreciation, rather than his reaction to the particular programme, may be the explanation for the above findings. A good deal of further detailed research is needed here.

8.6 'Coronation Street': a case history

An example of how these kinds of result can already help to explain practical findings is provided by results in 1972 for the twice-weekly ITV programme 'Coronation Street'.

In a certain analysis over several months it was found that the AI scores for 'Coronation Street' were consistently higher on Wednesdays than on Mondays:

average Monday AI	70
average Wednesday AI	73

Why did two adjacent episodes of the same long-running programme register such a consistent (if small) difference in audience appreciation?

The situation appears at first sight more curious when we take audience size into account. The Monday episodes were seen consistently by fractionally more people than the Wednesday ones:

average Monday rating	40
average Wednesday rating	38

Thus, despite somewhat lower AI scores given to the Monday episodes, slightly more people saw them than those on Wednesdays.

The explanation lies in the fact that some viewers saw both 'Coronation Street' episodes in a given week ('repeat-viewers') and some only one ('new' or 'lapsed' viewers). In line with the more general findings illustrated in Table 8.7 Monday–Wednesday repeat-viewers gave higher AI scores than the others, the difference being here as much as 15 AI points:

average AI score of repeat-viewers	75
average AI score of one-episode-only viewers	60

The repeat-viewers gave almost the same average scores to the Monday and Wednesday episodes and, by definition, the number of repeat-viewers is the same on Monday and on Wednesday – they are the same people. But since the rating of the Monday episode was higher than that of the Wednesday one, there were more one-episode-only viewers on Monday, i.e. more of those who tend to give the programme lower AI scores. The Monday audience therefore contained more relatively low AI-scoring viewers than the Wednesday audience. The difference in the AI scores of the Monday and Wednesday episodes was therefore not due to any difference in the intrinsic merit of the episodes, nor how the same people evaluated them. Instead it was due to whatever other factors (e.g. viewing habits, opposing programmes, etc.) caused the Monday episode to be seen by slightly more viewers than that shown on Wednesday.

8.7 Interesting or enjoyable?

We have remarked above on the probability that different types of programmes – especially 'serious' versus 'light' programmes – will be judged against different criteria. This is acknowledged in the AURA system by the simple expedient of asking people to rate a programme as being 'interesting and/or enjoyable'. This approach was tested in a split-sample experiment in 1981. One half of the sample was asked to rate a number of programmes that they had seen on both the standard AI scale and a scale mentioning only 'interest': the 'Interesting Index' or II. The other half was asked to rate programmes on both the standard AI scale and a scale mentioning only 'enjoyment': the 'Enjoyment Index' or EI.

The three indices gave broadly similar results – for example, average scores of about 70. To this extent the three scales seemed mostly to reflect people's general 'liking' of the programmes they watched. But there were some systematic differences between serious or 'information' programmes and lighter 'entertainment' programmes. These differences are summarised in Table 8.8.

'Information' programmes were scored on average almost 10 points higher on the interesting scale (II) than on the enjoyable scale (EI). This is shown in the II and EI columns of the table. For 'entertainment' programmes there was a difference in the opposite direction but it was much smaller.

The major finding, however, was that the standard AI score was generally close to the higher of the II or EI scores. Thus, for the more demanding information programmes the average AI (74) in Table 8.8 equals the average II (74) rather than the average EI (66). For entertainment programmes the average AI (72) is slightly closer to the average EI (71) than to the average II (69), even though the EI versus II differences are small.

The implication is that the standard 'composite' AI scale reflects the viewer's more positive reaction, i.e. how well he or she likes the programme for being *either* interesting *or* enjoyable. This is the judgment the AI seeks.

Table 8.8
Programme type averages

Programme Type	(IBA Code)	II	EI	AI
'Information' Programmes				
Reg. news & c. aff.	(5)	74	64	74
Gen. int. & inf.	(6)	74	67	74
Other news	(5)	74	66	75
Average (3 types)		74	66	74
'Entertainment' Programmes				
Romance	(2)	68	71	71
Comedy & L.E.	(4)	68	71	71
Films & plays	(7)	68	70	70
Adv./action	(1)	70	70	71
Sport	(3)	72	73	75
Average (5 types)		69	71	72

8.8 Summary

The level of viewers' expressed 'appreciation' of the programmes they have seen varies with the size of the audience, but not in a simple way. For programme types the relationship is inverse: programme types with smaller audiences (usually more 'serious' programmes) tend to achieve rather higher appreciation. Perhaps the viewer demands more satisfaction before watching it (or watching enough of a serious programme to express an opinion about it). The programme is more demanding of the viewer, who in turn expects a greater return for the added 'investment'.

Within a programme type the tendency is for higher appreciation to go with higher ratings. Programmes that are seen by more viewers are more appreciated by those viewers on average. The effect, however, is not strong. As the rating goes up by 10 points, the AI rises by only 3 or 4 points. And there is considerable residual scatter for individual programmes. Overall, the size of a programme's audience is a rather poor predictor of the satisfaction it gives those who view it.

Nor do appreciation scores correlate clearly with the level of repeat-viewing for different programmes. This is probably because a programme's overall repeat-viewing level does not itself appear to imply any special liking or disliking of the programme.

Frequent viewers of a repetitive programme do, however, tend to give higher appreciation scores than infrequent viewers. But this may be at least in part because heavy viewers of television generally 'like' all programmes more.

It is clear from the various cases discussed in this chapter that viewers' attitudes towards television programming can only be properly interpreted in the context of their actual viewing behaviour.

9 The Liking of Programme Types

The studies of viewing behaviour summarised in earlier chapters have consistently shown that there is little or no 'programme type' effect, whereby different programmes of the same type attract the same group of viewers. Nonetheless, one feels that two adventure series programmes like 'Hawaii Five-O' and 'The Persuaders', should have some common appeal. Some people must like various programmes of this sort, some must prefer light entertainment programmes, others sport and so on.

In fact there is evidence that this is so in terms of the programmes people say they like to watch. But this occurrence of programme-type clusters in terms of what people say they like can also be reconciled with the absence of such clusters in terms of what people actually do.

The data analysed here come from a question in the 1972 Leo Burnett *Life Style Research* study (see Segnit and Broadbent, 1971). Some 7000 adults were asked to say for each of 55 ITV or BBC programmes whether they:

'Really like to watch it';
'Watch it only because someone in my family likes it';
'Watch it when there's nothing better'; or
'Don't watch it'.

It is important to note that 'watching' a programme here refers to some general tendency to watch that programme sometimes and not necessarily to having watched it on its last screening.

9.1 Programme clusters

To illustrate we start with two marked groupings or clusters – sports programmes and current affairs programmes – which show up in terms of what people said they 'really like to watch'. For simplicity we shall refer to this as 'liking'.

Table 9.1 shows the percentage of adult viewers who 'liked' one sports programme and also 'liked' each of the other four sports programmes analysed. The table format is that of the duplication of viewing tables in earlier chapters and the results also are similar. Thus, there is little variation of the figures in each column from their average. 'World of Sport' was liked by about 76% of those who liked any of the other sports programmes and 'Rugby Special' was liked by about 30%.

These averages are all about twice as high as the percentages among adults as a whole who 'like' the programmes. This 2:1 ratio applies for each separate programme.

Table 9.1

Sports programmes

(% of adults saying they 'really like to watch' one programme who also say they 'really like to watch' another)

UK Adults 1972		**Who also like to watch**				
		World of Sport	Match of the Day	Grand- stand	Prof. Boxing	Rugby Special
Adults who like to watch						
ITV World of Sport	100%	–	73	72	61	28
BBC Match of the Day	100%	75	–	71	60	30
BBC Grandstand	100%	80	77	–	62	32
ITV Prof. Boxing	100%	75	72	68	–	31
BBC Rugby Special	100%	74	75	75	65	–
Average	100%	76	74	72	62	30
ALL ADULTS	**100%**	**39**	**38**	**35**	**32**	**15**

Table 9.2 shows a similar pattern for five current affairs programmes. The percentage liking a programme amongst 'likers' of another current affairs programme is again almost double the percentage liking the programme in the population as a whole.

These high 'likings' could merely represent a general tendency amongst 'likers' of one programme to 'like' other programmes, irrespective of their type. Table 9.3 shows that this is partly so but not the full explanation. The current affairs programmes were liked substantially less often by likers of the sports programmes. There are real groupings here by programme type.

Table 9.4 summarises these results. In the last column we see that the average current affairs programme is liked by 46% of the likers of the other current affairs programmes and by only 31% of the likers of the sports programmes. But among adults as a whole the figure is even lower, at 24%.

There is therefore some tendency for likers of one type of programme also to have a more than average liking for a programme of another type (the 31%

Table 9.2
Current affairs programmes

UK Adults 1972		**Who also like to watch**				
		Panorama	24 Hours	This Week	Today	Line Up
Adults who like to watch						
BBC Panorama	100%	–	68	50	37	18
BBC 24 Hours	100%	67	–	53	39	20
ITV This Week	100%	58	61	–	43	19
ITV Today	100%	47	50	48	–	17
BBC Line Up	100%	59	66	53	44	–
Average	100%	58	61	51	41	19
ALL ADULTS	**100%**	**31**	**30**	**27**	**24**	**9**

Table 9.3
Sports versus current affairs

UK Adults 1972		**Who also like to watch**				
		Panorama	24 Hours	This Week	Today	Line Up
Adults who like to watch						
World of Sport	100%	41	39	34	29	12
Match of the Day	100%	38	38	31	26	11
Grandstand	100%	42	40	34	28	12
Prof. Boxing	100%	42	39	34	28	13
Rugby Special	100%	47	44	33	29	16
Average	100%	42	40	33	28	13
ALL ADULTS	**100%**	**31**	**30**	**27**	**24**	**9**

versus 24% of the previous paragraph). This occurs generally. It seems to reflect the existence of heavier viewers who tend to watch – and often to 'like' – a wide range of programmes. There is certainly almost no case of fewer likers of one programme type liking another (i.e. that fewer like it than occurs among all adults). The special clusters of more intense correlations for programmes of the same type, illustrated in Tables 9.1 and 9.2, are therefore superimposed on this general tendency for some people to like TV programmes generally.

Table 9.5 gives another illustration for five light entertainment programmes on ITV, contrasted with adventure series and other ITV programmes. The contrast is clear.

Table 9.4
Likers of the current affairs programmes

UK Adults 1972	*% who like to watch*					
	Panorama	24 Hours	This Week	Today	Line Up	**Average prog**
Amongst likers of the						
– OTHER C.A. progs	58	61	51	41	19	**46**
– Sports programmes	42	40	33	28	13	**31**
ALL ADULTS	**31**	**30**	**27**	**24**	**9**	**24**

Table 9.5
Liking of light entertainment on ITV

UK Adults 1972	*Average % who like to watch*					
	Opportun-ity Knocks	Family at War	Corona-tion Street	Golden Shot	Peyton Place	**Average**
Amongst likers of the						
– OTHER L.E. progs	68	61	59	49	31	**54**
– Adventure Series	52	51	44	37	26	**42**
– OTHER ITV progs	52	45	40	38	21	**39**
ALL ADULTS	**42**	**40**	**34**	**28**	**18**	**32**

Six programme clusters

Altogether, only six recognisable programme clusters have emerged from the data analysed so far (55 programmes covered in the Leo Burnett survey). They are summarised, with typical 1972 programmes, in Table 9.6.

It is important to stress that these clusters have been defined by noting the cases where people who say they 'really like to watch' one particular programme include an especially high proportion of people who say they 'really like to watch' another. The groupings have not come from any direct assessment of programme content or treatment. Nonetheless, five of the clusters agree with 'commonsense' programme-type classifications (as also formalised by the IBA for example). These clusters can therefore be readily named, as is done in Table 9.6.

Only the sixth cluster does not fit into an existing category. (One can imagine these programmes being liked by the same people but it is not clear how to say in standard terms what the programmes have in common; Leo Burnett refers to such programmes as 'cult' programmes.)

The typology in Table 9.6 should not be overinterpreted. It does not mean that viewers can be divided into separate sub-groups who like one type of programme and not others. It is not an exclusive classification in that sense. Instead, it only reflects groupings of certain above-average likings.

Table 9.6
Six programme clusters
(The six main programme groups emerging from the analysis)

Programme Cluster	Examples on ITV	Examples on BBC
1. *SPORTS*	World of Sport Professional Boxing	Grandstand Match of the Day Rugby Special
2. *CURRENT AFFAIRS*	Today This Week Aquarius	Late Night Line Up Talk Back 24 Hours Panorama
3. *LIGHT ENTERTAINMENT* (a) Serials	(a) Coronation St. Peyton Place Family at War	
(b) General	(b) Opportunity Knocks Golden Shot This is Your Life Mike & Bernie	(b) Z-Cars Owen M.D. Galloping Gourmet
(c) Sit. Comedy	(c) Please Sir On the Buses	(c) Now Take My Wife Here's Lucy
4. *ADVENTURE*	Public Eye Callan Jason King Hawaii Five-O	Ironside The Virginian
5. *CHILDREN'S*		Magic Roundabout Blue Peter
6. (*Not Named*)	Thunderbirds	Star Trek Pink Panther Monty Python Top of the Pops

Exceptions

The analysis here is based on only one set of programmes at one particular point in time (1972) and further work is needed. (Corresponding analyses of the AI type of data which was described in Chapter 8 are leading to broadly similar results. The AI data are, however, complicated for this particular purpose by the assessments being related to the programmes actually seen that week.)

Some exceptions to the main clusters are worth noting at this early stage. A few programmes among the 55, like the highly popular comedy show 'Morecambe and Wise', could not readily be assigned to any programme cluster. One or two other programmes fell into more than one group. 'Wrestling' on ITV is one such example. As shown in Table 9.7 it is popular among those who said they liked to watch ITV's 'World of Sport' and 'Boxing'. But it is less popular among those who said they liked BBC's sports programmes 'Match of the Day' and 'Rugby Special'. In fact, 'Wrestling' is more popular with people who said they liked entertainment programmes like 'Golden Shot' or 'Opportunity Knocks'. 'Wrestling' therefore appears to appeal to two somewhat different groups of people for two different reasons – to one it is sport, to the other light entertainment.

Table 9.7
Wrestling
(% of adults saying they 'really like to watch' each programme
who also say they 'really like to watch' wrestling)

UK Adults 1972		*Who also like to watch* Wrestling
Adults who like to watch		
World of Sport	100%	48 ⎱ 50
Prof. Boxing	100%	51 ⎰
Match of the Day	100%	40 ⎫
Grandstand	100%	43 ⎬ 40
Rugby Special	100%	37 ⎭
Golden Shot	100%	49 ⎱ 47
Opportunity Knocks	100%	46 ⎰
ALL ADULTS	**100%**	**32**

9.2 Programme character

These data on the programmes which viewers say they really like to watch can be used to define the appeal or 'character' of a programme. This can be done by noting how popular the programme is among the people who like various other programmes. The approach is illustrated in Table 9.8 for 'Family at War'.

Table 9.8
Programme character: 'Family at War'
(10 programmes whose 'likers' most like or least like 'Family at War')

UK Adults 1972	% who say they like to watch FAMILY AT WAR	
ALL ADULTS	**40**	
Among adults who like		*High*
Peyton Place	72	
Coronation Street	67	
This is Your Life	58	
Golden Shot	55	
Opportunity Knocks	54	
Mike & Bernie	52	
Here's Lucy	52	
Owen M.D.	52	
Public Eye	52	
Jason King	53	
		Low
Pink Panther	32	
Monty Python	30	
Magic Roundabout	32	
Panorama	33	
Tomorrow's World	31	
World of Sport	33	
Grandstand	30	
Match of the Day	29	
Prof. Boxing	28	
Rugby Special	26	

From the 54 other programmes covered, the 10 programmes are shown whose fans include the highest percentage of people also liking 'Family at War' and the 10 programmes which have the least likers in common with it. Thus, across the population as a whole 40% say they like 'Family at War'. But the people who like other programmes show a great deal of variation in this respect, from a high of 72% to a low of 26%.

At one extreme, 'Family at War' is very popular with people who like the other two ITV serials, 'Peyton Place' and 'Coronation Street' (being liked by about 70% of these). It is also relatively popular with people who said that they liked other light entertainment programmes and with those who said they liked adventure series such as 'Public Eye' and 'Jason King'.

At the other extreme, relatively few of those who said they liked current affairs, sports or the 'Monty Python' group of programmes (cluster 6 in Table 9.6) also said they liked 'Family at War'. Here the incidence of people saying they liked 'Family at War' is actually well below the level among all adults – one of the quite rare instances of this in the current data. The result for 'Family at War' is therefore a wide spectrum of opinions.

Table 9.9 gives a slightly more complicated example, for 'Aquarius'. The general level of liking is much lower (only 12% of adults as a whole do so). 'Aquarius' is relatively popular among those who said they liked some of the current affairs programmes, especially 'Line Up', but not particularly among fans of other programmes in the group such as 'Panorama'. 'Aquarius' also appeals to people who said they liked the 'Monty Python' and adventure types of programme. However, the programme is relatively unpopular with both extremes of the dimension seen for 'Family at War' – the fans of 'This is Your Life' or 'Coronation Street' and the fans of the sports programmes.

The usefulnes of such analyses will depend on experience of a wide range of examples and the development of a 'feel' for different patterns of preferences. In particular, it will be possible to differentiate in detailed terms between some programmes which are in most respects similar. For instance, 'Talk Back' and 'Line Up' had a particularly high overlap in terms of the people who said they liked them. Both were popular among people who said they liked other current affairs programmes. But 'Talk Back' was also relatively popular with those who said they liked some of the lighter programmes such as 'Galloping Gourmet' or 'Owen M.D.', while 'Line Up' was popular with those who said they liked more 'intellectual' programmes such as 'Aquarius' and 'Monty Python'.

9.3 The nature of programme clusters

The programme clusters described in Table 9.6 are not unexpected on common-sense grounds. One expects certain people to like programmes of a certain type. Nonetheless, the nature of the clusters does not necessarily turn out to be self-evident when we analyse them further.

Table 9.9
Programme character: 'Aquarius'
(10 programmes whose 'likers' most like, or least like
'Aquarius')

UK Adults 1972	% who say they like to watch AQUARIUS	
ALL ADULTS	**12**	
Among adults who like:		*High*
Line Up	28	
This Week	20	
Parkinson	17	
Braden's Week	17	
Monty Python	19	
Thunderbirds	19	
Star Trek	17	
Jason King	17	
Callan	17	
The X Film	17	
		Low
Owen M.D.	12	
This is Your Life	11	
Coronation Street	11	
Val Doonican	11	
Golden Shot	11	
Generation Game	11	
World of Sport	12	
Grandstand	11	
Rugby Special	11	
Match of the Day	10	

Sports and light entertainment programmes are generally known to have markedly different appeal to two different groups, men and women. Thus, the five sports programmes in Table 9.1 are typically more popular with men, almost 50% of whom said they liked these programmes compared with only 16% of women. In contrast, light entertainment programmes are generally more popular among women, 42% of whom liked the average programme in Table 9.5 compared with only 22% of men.

Thus it might be thought that the sports cluster would be due to men (who like sports programmes) and the light entertainment cluster due to women.

Table 9.10
Sports: women and men
(% of women or men saying they 'really like to watch' one sports programme who also
'really like to watch' other sports programmes)

UK 1972		World of Sport	Match of the Day	Grand-stand	Prof. Boxing	Rugby Special
			Who also like to watch			
WOMEN who like to watch						
World of Sport	100%	–	57	57	35	23
Match of the Day	100%	67	–	57	33	24
Grandstand	100%	72	61	–	32	27
Prof. Boxing	100%	66	52	48	–	27
Rugby Special	100%	65	58	61	41	–
Av. Sports Progr.	100%	68	57	56	35	25
ALL WOMEN	**100%**	**23**	**29**	**18**	**12**	**8**
MEN who like to watch						
World of Sport	100%	–	80	78	73	31
Match of the Day	100%	78	–	76	70	32
Grandstand	100%	83	83	–	72	34
Prof. Boxing	100%	77	77	73	–	32
Rugby Special	100%	77	81	81	75	–
Av. Sports Progr.	100%	79	80	77	73	32
ALL MEN	**100%**	**57**	**59**	**54**	**54**	**23**

But the reverse is the case. This can be seen from Table 9.10 which shows separately for women and for men the percentages liking sports programmes among those liking the other four sports programmes, together with the percentage of all women or all men who liked each of the sports programmes.

For women there is a ratio of about 3:1 for each sports programme between likers of other sports programmes and all women. In contrast, the corresponding differences for men are much smaller; the ratio of the figures in the last two rows of Table 9.10 is only about 1.4:1.

It follows that there is only a small group of women who like sports programmes and that is where the clustering of sports programmes among all adults mainly stems from. The men themselves only show relatively weak clustering (since so many like the sports programmes anyway).

The same pattern holds for light entertainment programmes but with the role of men and women reversed. This is summarised in Table 9.11. Here there is relatively little clustering of the programmes among women, the differences between likers and all women being relatively small (the averages of 59% and 44%, a ratio of only 1.3:1). In contrast, only an average of 22%

of all men liked the ITV light entertainment programmes whereas 43% of the male likers did so – nearly twice as many.

Table 9.11
Light entertainment: women and men
(the average % of women or men saying they 'really like to watch' one light entertainment programme who also said they 'really like to watch' another)

UK 1972		Oppor-tunity Knocks	Family at War	Coro-nation Street	Golden Shot	Peyton Place	Aver-age
		Who also like to watch					
WOMEN who like to watch							
Av. OTHER L.E.							
progr.	100%	68	70	66	51	40	59
ALL WOMEN	**100%**	**46**	**55**	**46**	**44**	**28**	**44**
MEN who like to watch							
Av. OTHER L.E.							
progr.	100%	67	43	47	46	14	43
ALL MEN	**100%**	**37**	**23**	**21**	**24**	**7**	**22**

These results indicate that the apparent clustering of programme preferences noted in this chapter is more a negative than a positive effect. It may reflect the existence of a group who do not like a particular group of programmes. More work is needed in this area.

9.4 Reconciliation with viewing behaviour

Whatever its underlying nature, the clustering of viewers' liking for programmes of the same type seems at first sight to contradict the finding that no such programme-type clusters have emerged for people's actual viewing behaviour (Chapter 4). But in fact the two kinds of results are not inconsistent.

People often watch programmes which they would not claim to like particularly. Apart from mentioning programmes which they 'really like to watch', people in the Leo Burnett survey analysed in this chapter were also asked to mention programmes which they watched because 'someone in the family likes them' or because 'there's nothing better to watch'. For example, only 59% of adults claimed to watch 'Panorama' at all. Of these, 31% said they really liked to watch it, 8% said they watched it because someone in the family liked it and 15% said they watched it because there was nothing better to watch. The 'enforced' viewing of 'Panorama' by 23% out of the 59% who view serves to dilute the effect of programme preferences on actual viewing behaviour.

Table 9.12
Preference versus viewing behaviour: current affairs
(% who either said they 'really like to watch' or 'view for any reason' each current affairs programme among those who also said they 'really like to watch' or 'view for any reason': OTHER current affairs – averaged over 4 programmes; and sports – averaged over 5 programmes)

UK Adults 1972		Who also 'really like to watch'					
		Pano-rama	24 Hours	This Week	To-day	Line Up	Av. C.A.
Adults who really like to watch							
Av. OTHER C.A.	100%	58	61	51	41	19	46
Av. Sports	100%	42	40	33	28	13	31

		Who also 'watch for any reason'					
		Pano-rama	24 Hours	This Week	To-day	Line Up	Av. C.A.
Adults who watch for any reason							
Av. OTHER C.A.	100%	74	78	67	51	32	60
Av. Sport	100%	64	66	58	45	28	52

This dilution is illustrated in Table 9.12 for the current affairs versus sports programmes analysed earlier. The top part of the table repeats from Table 9.4 the average percentages who 'really like to watch' each current affairs programme, both among those people who 'really like to watch' another current affairs programme (an average of 46%) and among those who 'really like to watch' a sports programme (an average of 31%). The ratio is about 1.5:1.

The bottom part of the table gives comparable figures relating to people's total viewing claims, i.e. including 'enforced' viewing of both kinds. It shows the average extent to which each current affairs programme was said to be watched for any reason by the people who for any reason watched another current affairs programme (an average of 60%) and by those who for any reason watched a sports programme (an average of 52%). The figure for duplicated viewing within the current affairs group is still somewhat the higher but only by relatively little – an average ratio of just over 1.1:1. The clustering of different current affairs programmes is therefore very much weaker in terms of something which resembles actual measured viewing (i.e. 'watching for any reason') rather than claims to like watching.

This is a general finding. Table 9.13 sets out the corresponding (but fuller) data for total viewing claims for two other programme groups, light entertainments and adventure series. A light entertainment programme is watched (for any reason) by on average 69% of those who watched (for any reason) the other light entertainments and by on average 66% of those who watched (for any reason) an adventure series – the figures are almost the same. The difference which typically occurred for 'liking' claims (e.g. an average of 54% among likers of the light entertainment programmes versus 42% among likers of the average adventure programme in Table 9.5) has virtually disappeared here in terms of people's general viewing behaviour.

An additional factor in reconciling the liking and viewing results is that in any particular week some people tend to miss seeing even those programmes which they particularly like. Viewing claims made in the context of the Leo Burnett survey analysed here are claims to view a given programme sometimes. The total claims to watch a programme 'for any reason', e.g. the 59% claim for 'Panorama', are well above the actual ratings normally achieved by any single

Table 9.13
ITV light entertainment and adventure series: viewing for any reason
(% of adults claiming to view one programme for any reason who also claim to view another for any reason)

		Who also view (for any reason)				
UK Adults *1972*		Opportunity Knocks	Family at War	Coronation Street	Golden Shot	Peyton Place
Adults who view (for any reason)						
Opportunity Knocks	100%	–	68	71	82	39
Family at War	100%	79	–	72	77	42
Coronation Street	100%	86	75	–	81	43
Golden Shot	100%	82	66	67	–	37
Peyton Place	100%	85	80	79	83	–
Av. LIGHT ENTERTAINMENT		**83**	**72**	**72**	**80**	**40**
Thunderbirds	100%	79	70	66	80	45
Hawaii Five-O	100%	76	70	65	76	40
Jason King	100%	80	73	71	79	43
Persuaders	100%	76	69	65	75	38
Callan	100%	75	70	65	75	40
Public Eye	100%	78	73	68	77	41
Av. ADVENTURE SERIES		**76**	**71**	**66**	**77**	**41**

showing of the programme. This is in line with what we already know about the irregularity of viewing of a programme series. Only just over half of those who see one given episode of a programme also see another given episode of it; hence many more see it sometimes during a period of weeks than see a single episode.

This irregularity of actual viewing behaviour (together with a certain amount of false claiming which can occur in an interview situation) could appear sufficient to complete the reconciliation of quite marked programme-type patterns in claimed viewing preferences with the lack of such patterns in the general viewing data.

9.5 Summary

Programmes fall into a recognisable classification in terms of what people say they 'really like to watch'. From some early analyses five named programme types have emerged so far: sports, current affairs, light entertainment, adventure and children's programmes. Programmes fall into such groups in so far as the people who say they like one programme in the group include an especially high proportion of people who also say they like other programmes in the same group. This classification of programmes is not exclusive: people who say they particularly like to watch one type of programme do not, in general, say they dislike other types.

However, the way in which these programme-type clusters appear to arise is perhaps unexpected. The underlying factor seems to be the existence of a group of people who do not like a particular type of programme.

The programme-type patterns in terms of claimed programme preferences have not been found in studies of actual viewing behaviour. This is partly due to the fact that a considerable part of a person's viewing is 'enforced' – watching programmes because other family members want to see them or simply because there is nothing better to watch – and partly because of irregular viewing (only about 55% of the audience to one episode of a programme will watch the same programme in another week). Many people do not watch regularly even those programmes which they say they really like to watch.

10 Television as a Medium

In this final chapter we shall draw together the findings described in this book and briefly comment on some of the broader implications for television as a medium.

10.1 Simple findings

The results in this book are simple. There is one main result concerning the repeat-viewing of a given programme, one main result concerning the duplication between the audiences of different programmes and some findings concerning the audience's expressed appreciation or liking of a programme. And the crux of this second edition is that some 10 years after the first these same results still hold.

Repeat-viewing

The basic finding here is that about 55% of viewers of one episode of a programme also watch the following episode (usually a week later) for all those cases where the ratings are more or less steady from episode to episode. This single figure – 55% – summarises the main result succinctly.

Repeat-viewing levels, however, decrease somewhat with audience size – the smaller the audience, the lower the incidence of repeat-viewers. But otherwise repeat-viewing levels of different programmes vary relatively little, although exceptions occur. In particular, there is virtually no systematic variation by programme type or content. Repeat-viewing of a serial with a continuing story-line is generally no higher than that for a regular film slot with radically different showings each week.

Nor is there evidence of any marked 'erosion' in the degree of repeat-viewing for episodes shown more than one week apart in cases where, once more, rating levels are steady. Failure to repeat-view seems to be a reflection of irregular or infrequent viewing habits, not of any special dislike or lack of interest in what has already been seen. This conclusion is supported by the finding that repeat-viewing is about 50% even for irregular programming, i.e. when altogether different programmes (with similar ratings) are shown in the same time-slot each week. Repeat-viewing therefore appears to be more a function of social habits (i.e. people's availability) than of programme content.

Extension of the analyses to more than two episodes of a programme also shows regular results and the development of theory. For any extended series of episodes the main finding is that almost no one sees all, or even nearly all, the different episodes.

Audience duplication

The extent to which different programmes share the same viewers also follows a simple pattern. This is expressed by the duplication of viewing law. Thus, for any two programmes the level of duplication in their audiences depends on the ratings of the programmes and not on their content. One pair of programmes generally has the same degree of audience duplication as any other pair of programmes with the same ratings.

There are no sub-patterns in this by programme type but there are differences by channel. Audience duplication is somewhat higher for two programmes on the same channel than for ones on different channels, although the effect is not very big. This is 'channel loyalty' – some people are consistently heavier viewers of one channel than another.

The theoretical explanation of the duplication of viewing law is in terms of people's differing patterns of viewing and programme preference. One person watches programmes A, B and C, another watches A, X and Y, a third B, Y and M and so on. It is not a case of there being large sub-groups of people with common viewing patterns (other than is reflected in the sheer audience size for different programmes – the ratings). The observed patterns of audience duplication derive essentially from people's individualistic use of the medium, coupled with marked differences in their total amounts of viewing.

For two programmes both shown in the afternoon or in the early evening of two different week-days, audience duplication levels are rather high in relation to the ratings of these programmes. This is, however, largely a matter of non-availability. The same people tend to be out on different week-day afternoons. Audience overlap is therefore high compared with the relatively low ratings in the population as a whole. The same occurs for pairs of programmes shown late on two different evenings. If allowance for regular availability is made (e.g. the same people tending to stay up fairly late), the ratio of duplication to rating levels becomes more normal.

Another consistent sub-pattern is that successive programmes on a channel have higher duplicated audiences. But this 'inheritance' effect applies only to the adjacent programme and, to a lesser extent, to the next but one programme. It does not apply throughout the evening.

Audience appreciation

When people are asked about their liking or appreciation of programmes viewed in a given week, viewers' average 'appreciation score' does not depend on the rating of the programme or on the incidence of repeat-viewing. But people giving a high appreciation to a particular programme are more likely to see it often.

When people say they 'really like to watch' a certain programme, they tend to say this also about other programmes of the same type (e.g. sport). This clustering of programmes people say they like to watch reconciles with the lack of such programme-type preferences in actual viewing behaviour, partly because people do not always watch even programmes they really like and partly because they also watch programmes other than ones they 'really like to watch' (i.e. if 'someone else in the family likes it' or if 'nothing better was on').

10.2 Some broad implications

The main implication of these various findings is that television as currently operated is indeed a mass medium. Instead of being complex, with much differentiation between distinct groups of viewers or between the audiences of different programmes, viewing behaviour and audience appreciation appear to follow a few very general and simple patterns operating right across the board.

For example, it has been long established that the AB social classes in the UK view somewhat less television than does the public as a whole. Nonetheless, the actual distributions of viewing times between different programmes are similar. This is so even though preference is expressed by the ABs for more 'serious' material (e.g. Marplan, 1965). But there are discrepancies between what viewers say or feel they would like to watch and what they watch in practice. A similar conclusion was reached in Steiner's (1963) study of viewing tastes and behaviour in the US, reinforced by a more recent follow-up (Bower, 1973).

The patterns of viewing behaviour established in this book cannot in themselves determine the decisions that ought to be made, either about television in general or about particular programmes. For decisions, certain social or communication criteria or targets need to be set. But in as far as these refer to the viewer, such criteria can now be informed by, and evaluated against, our knowledge of viewer behaviour. (For example, we now know that it would be absurd to set a target of 90% repeat-viewers for successive episodes of a new series – that sort of thing simply does not happen. And if we observe a repeat-level of about 50 to 60%, we now know that this is normal. It also follows that we cannot expect many people to see all or most episodes in any programme series. The medium does not work like that.)

The results in this book would not in themselves, for example, have said in the seventies whether or not there should be a fourth channel in the UK, how it ought to be organised or what sort of programming policy ought to be adopted. Some kind of target – like 'increasing viewers' choice of programmes' – had to be set. But we can now consider whether provision of an additional channel really results in increased choice or what such a target means anyway. Viewers may not use the additional 'freedom of choice' appropriately.

Thus, the evidence is that most viewers have a wide spread of interests and desires as far as their viewing is concerned. This suggests that individual needs might perhaps most effectively be met through the widest possible choice of programmes. But with few channels, minority interests cannot be met for more than a fraction of the time if majority tastes are to be catered for most of the time. And if the channels directly compete for viewers, minority interests will hardly be served at all: each channel will compete for the majority audience through similar offerings, as tends to occur on the national networks in the US.

A common argument is therefore that given a limited number of channels, viewer satisfaction should be maximised by complementary programming (e.g. Steiner, 1963). At any given time, the channels would be showing programmes which appeal to widely differing interest groups. (The programme choice available could also grow with any increased use of new technologies: VCRs, DBS, cable television, etc. – potentially yielding vastly more channels.)

The basic question then is what actual use the public makes of the greater variety of choice that might be or is on offer. Would complementary programming, with some items catering for specialist groups, be regarded as failing if students of literature were still to relax with 'The Mary Tyler Moore Show' rather than watch 'The White Devil' on the other channel, i.e. if even the minority audiences were largely to evaporate? This has been shown to occur to quite an extent by BBC2 in the UK and the public service channels in the US. 'Minority' programmes are often not greatly watched even by their supposedly relevant minority if entertainment programmes are available simultaneously. This does not mean that no programme with obvious minority appeal should be screened but that any programme decisions should be made whilst knowing what size and kind of viewing audience to expect. And a small rating in the UK of say 1 or 2 is still a million viewers!

A related problem concerns the broad programming policies of authorities like the IBA or BBC. The objective cannot be to maximise audience size in general. Individual programme companies and programme planners may already be trying to do that and in any case it is not regarded as socially proper always to try for maximum audiences. But obviously the target cannot be to minimise audiences.

So an alternative target is set – balance. But we are now learning more of what balance in programming might actually mean to the viewer. It does not necessarily mean limiting the number of westerns that may be shown in any one week, implying that these would merely tend to cater to some groups of western addicts. There appear to be no such addicts. People who watch one particular western are no more likely to watch other westerns than are other viewers.

If an additional western were screened by ITV and achieved a rating of 20, then about 34% (i.e. 1.7 × 20) of the audience of any other ITV programme (whether another western or anything else) would watch the new western but some 66% would not. And about 18% (i.e. 0.9 × 20) of the audience of any BBC1 programme would watch the extra western on ITV but 80% would not. Much the same numbers would (or would not) watch any other ITV programme with a rating of 20.

Balance is also a widely used concept for controversial matters – political or other. But 'putting the other side' in another broadcast might not seem very effective if few of those who were exposed to the first broadcast saw the second (or vice versa). Yet this is what generally happens – about 55% if it is exactly a week later and generally less if it is on another day of the same week. (The main exceptions are election and party political broadcasts in the UK which are generally shown simultaneously on all channels. They therefore achieve relatively high total ratings and correspondingly high duplication – heavy viewers do not switch off – although rather low appreciation scores.)

The conclusion for handling controversial matters might therefore be always to put both sides in the same broadcast. But that would often be retrogressive in terms of creative broadcasting and possibly insulting to the audience. People are not always that easily influenced or manipulated by a single broadcast. (The evidence for election broadcasts is that they neither preach to the converted nor do the opposite – thus a Labour Party broadcast is not much more likely to be seen by viewers of other Labour broadcasts than by viewers of Conservative or Liberal Party broadcasts.)

One of the simplest lessons of this book concerns people's low exposure to complete series of broadcasts. Producers, critics and the public need to learn that very few viewers of any series see all or most of its episodes. A series of programmes may form a unity to the producer but not to the audience. It is a mistake to suppose that on television anyone ever 'reads the whole book through'. One may doubt whether Marshall McLuhan knew it this way, but television certainly is not a 'linear' medium.

10.3 The pull of the box

The finding that repeat-viewing levels are relatively low (55%) may seem difficult to square with the supposedly compulsive attraction of television. Surely a highly popular programme series will tend to be watched very regularly? It is true that some people will do so but they will be a small minority. The popularity or high rating of the series largely arises from all the other, mostly irregular, viewers whom it attracts.

And even the 'regular viewers' are often not all that regular. People may claim that they always watch their favourite programme. But they claim this only for one or two programmes and not for all the other twenty or so they see each week. And however regularly one thinks one watches a programme, one actually only watches it in those weeks when one does not happen to be away or doing something else. (Like the football addict who watches 'Match of the Day' religiously about once every three weeks.)

Television is a medium where one usually hardly misses – is hardly even aware of – what one does not see (except perhaps when people talk about it next day). The pull of the box certainly exists when the set is on. But it seems to snap if one leaves the room and shuts the door, even in the middle of a programme. Suddenly the plot and the characters no longer matter so much. If one misses a favourite programme how much does one really mind afterwards? How much does missing it reduce one's enjoyment when watching the next episode a week later? If one comes in half way through, how much effort does one really make to be filled in on what one has missed or to see a repeat? The general run of programmes does not seem to be regarded as 'serious' in that sense. Television is mostly a 'low involvement' medium.

Even for 'specialist' programmes (usually defined as such only if they have small ratings) the position is similar. They certainly do not attract more regular viewers. If anything, they are viewed more by heavy (and hence by definition, relatively indiscriminate or catholic) viewers and not so much by specialist or otherwise selective viewers. (Light viewers tend to watch the popular programmes more – that is why they are popular.)

The explanation seems to be that people with a real specialist interest do not generally feel a need to follow it on television. Artists do not feel they need to watch art programmes; knitters, knitting programmes; or business-men, business programmes (unless perhaps some friend – or enemy – happens to be performing). Specialists already know all that. Even religious people do not religiously watch all their programmes but go to church instead.

The physical attraction of the live screen does not mean that the programme material as such always exercises any great pull. (Repeat-viewing is not high and viewers of a certain programme do not necessarily watch others of the same type – the duplication of viewing law.) A viewer develops a habit and hence a certain liking for a particular programme (e.g. 'Ironside') and tends to watch it fairly regularly. But that does not mean that he feels the need to watch other almost identical-seeming programmes (e.g. 'Cannon'). And most viewers also watch many programmes other than their favourites or the ones they 'really like to watch', as we saw in Chapters 8 and 9.

10.4 The effects of television

Given the large amount of time many people spend watching television, it is not surprising that concern has been expressed about its effects, i.e. its possible influence on opinions and behaviour. Politicians, social observers, programme makers and advertisers have all been conscious – perhaps too conscious – of the alleged power of television as a medium. As stressed earlier, we ourselves have little specific expertise in this wider area of assessment. But some brief comments seem necessary here.

Considerable social research into the effects of television has taken place in the UK, the US and various other countries. Sociologists and others have examined the impact of heavy television viewing on children (e.g. Himmelweit, Oppenheim and Vince, 1958; Belson, 1967; Halloran, Brown and Chaney, 1970; US Department of Health, Education and Welfare, 1972) and the role of television in politics (e.g. Klapper, 1960; Blumler and McQuail, 1968; Halloran, 1970). The relation of television to such topics as violence, education, the arts and religion has also been studied (Halloran, 1970; ITA, 1970; Rubenstein *et al.*, 1972; US Department of Health, Education and Welfare, 1972; Liebert *et al.*, 1973; Howitt and Cumberbatch, 1975). In addition, a vast and diffuse body of literature has been generated by journalists, politicians and other writers, but little firm clarification seems to have resulted. This was the situation in 1975 at the time of the first edition of this book. With the second some 10 years later, nothing much has changed.

A typical issue that has been widely discussed is the possibly stimulating effect of violence on television. Here it has now been widely recognised that people do not rush out into the street to imitate what they have just seen on the screen. Any effects must be more diffused. There is a good deal of violence around and it is only in that broader context that any additional effects of television can be understood.

Yet it is certainly true that westerns and other fictional television programmes continually portray a world in which both good guys and bad guys use violence to solve problems and achieve goals. One might feel that the effects can hardly be wholesome. But it has also been argued to the contrary that such formalised violence on television might be cathartic – reducing rather than increasing tension in the viewer. There is no simple answer.

Television has of course also brought *real* violence into the home. We now almost daily 'see it how it is'. Apart, then, from the medium possibly acting as a stimulus to further violence (or to demonstrations which can lead to violence), it can also be argued that overexposure may have anaesthetised many of us. We may have grown used to seeing violence and its effects (or hunger and *its* effects). But would we actually act any differently day by day if we had not grown callous from seeing violence? Passing by on the other side of the street is at least as old as the Bible. So what we have come to take for granted is perhaps only violence on the screen and we may have become no more callous to the real thing than we already were. Indeed, as nations we may be becoming more concerned and socially conscious, perhaps because we are more informed and more aware. Typically, questions about the effects of television here are complex and clear-cut answers largely non-existent.

Trivialisation?

A different charge is that of trivialisation. With the advent of television, cinema attendances and radio listeners declined. It has also been suggested that theatre-going and reading of books decreased among heavy viewers (e.g. Belson, 1966). But how much of a loss was that? How "cultured" was Broadway or the West End theatre of London and how intellectual was the reading that was reduced? And what of the many who would not or did not go to the theatre or did not read? Do many of them now not occasionally see drama or serialisations of classics (even if sometimes only because they do not switch off)? Have drama groups and local theatres declined? And how do we allow for (and explain) the possibly greater financial support which society (government, business, the public) gives to the arts? Assessment and evaluation is again difficult.

Advertising

Another area is television advertising. Here the supposed power of television is more direct. The purpose is to influence specific attitudes and behaviour. Whatever the complexities of the advertising process may be, at least the aims and the means are relatively clear and explicit. Furthermore, there is a great deal of information directly relating to its expected effects – people's attitudes towards the advertised goods and their buying or consumption behaviour are widely measured. It is therefore possible to examine how advertising works through the media, as has been done more fully elsewhere (e.g. Ehrenberg, 1974). Here we briefly summarise the main arguments put forward. This also provides lessons on the power of television more generally.

We note that advertising, like television itself, is in an odd position. Its extreme protagonists claim it has extraordinary powers and its severest critics believe them. But whilst television advertising can be effective, it is not as powerful as is often thought. There is no evidence that it works by any strong form of persuasion or manipulation.

The possible effects of advertising on the demand for whole classes of goods or services (e.g. cars, cigarettes) must be distinguished from its effects on people's choice between competitive makes or brands of the product (e.g. Ford, Fiat or Volkswagen). Many of advertising's critics believe it has powers to create consumer demand for goods and to build our acquisitive society. But product-class advertising as a whole – 'Buy more cars', 'Drink more tea', etc. – cannot be held responsible. For one thing, there is relatively little of this form of advertising: for another, it generally produces only minor results. There are no dramatic claims in the literature (if we have missed one, that is the exception).

Repetitive advertising for individual brands – 'Buy Fords', 'Drink Tetley's Tea', etc. – is where the bulk of television advertising is concentrated. This could lead to a higher level of consumption of the product-class as a whole than would exist without it but there is no evidence that such secondary or even unintended effects (of brand advertising on total product-class demand) are either big or particularly common.

Whilst the bulk of advertising on television is for competitive brands, sales of these mostly do not actually change greatly from one year to the next. The great mass of brand advertising must therefore at best be defensive – the manufacturer aims to keep what he has by helping to reinforce an already-existing consumer habit of buying his brand. Any feeling of satisfaction with the brand – that it is liked at least no less than others available – has to be nurtured (with attitudes mostly changing after usage so as to reduce feelings of cognitive dissonance). People mostly ignore advertising for brands or products which they are not already using – it says little to them – by a process of selective perception.

But occasionally new customers for a brand or product are created, or altogether new brands or products are launched. Here advertising can both build awareness and help lead to an initial trial. The ultimate test is whether the consumer likes it after he has had it; only then will a repeat-buying habit (or favourable word-of-mouth recommendation) develop. The most widely quoted statistic is that nine out of ten new brands fail.

Advertising's role is seldom a very powerful one. It is not a matter of persuading or manipulating the ignorant consumer, since consumers of heavily advertised products are mostly highly experienced. They have usually already bought the product often and have used a wide range of different brands (Ehrenberg 1972). No exceptional liking or 'image' needs to be induced in the consumer because he knows similar brands to be similar and does not greatly care which he buys (which mainly matters to the manufacturer).

In its supposed role of creating our acquisitive society advertising tends to be confused with the influence of the mass media generally. Consumers' awareness and expectations have been raised by magazines, films and television, and also by people's greater mobility and education, not merely by advertising as such. Advertising itself generally follows changes in habits and fashion rather than leads. Many people want things which are hardly advertised at all.

Other effects

The way in which advertising appears to work must also tell us about other effects of television. Thus, the American President's National Commission on the Causes and Prevention of Violence in the late sixties argued that given that so many millions continue to be spent on television advertising to influence human behaviour, television advertising must be very powerful and hence television more generally must also have strong effects.

But if the effects of advertising itself are in fact relatively weak – mainly to reinforce those who are already using the brand and are already well informed and highly experienced, and with non-users of the brand or product hardly noticing the advertising – then we can expect the effects of the *programme* material on television to be even more diffuse. After all, the effects people worry about (like stimulating violence in the viewer, destroying his moral values or increasing his expectations as a consumer) are usually not even what the programme-makers are trying (or are paid) to sell.

Occasionally a television programme (or an advertisement) will leave viewers aware of something new (like a new form of violence or new attitudes towards others, or a new or previously ignored brand). But by itself this new awareness does not usually actually make one try the new thing, or give money to charity or be kind to foreigners, let alone will it inculcate a habit of doing so. The crucial factor is whether one likes the new thing after one has tried it. Rather than our attitudes causing behaviour, it is often the case that our more salient attitudes are affected by prior changes in our behaviour.

Any effect of television will tend to be slow and diffuse and difficult to isolate from the effects of other factors. The effects will take place in a wider context. Showing affluence or cruelty will mostly influence people who are ready to be influenced. There is more sex in real life than is ever shown on the screen, where it is most noticed by those looking for it.

We must not exaggerate what to expect from television, nor use inappropriate yardsticks and norms to evaluate it. Many readers of this book will be people who tend to feel guilty when watching television (or at least when admitting to it). But the mass of the population do not feel like that. For them, watching television is not necessarily regarded as a weak-minded substitute for doing something 'better'.

It is true that the fare which television provides is largely repetitive and often seems mindless. But when as young children we were read bedtime stories, we often asked for the same story (or even a certain favourite page) to be read over and over again and again. Less privileged children have no bedtime stories read to them at all.

Television provides information, entertainment and a way to pass the time in one's home. Much of it may be fairy stories but there is nothing new about that. Some television is good, most of it is mediocre and some poor. To understand it better and perhaps in some way to improve it, we must examine television from the point of view of the consumer. How do we, the viewers, use it? That is what this book has tried to illuminate. If in doing so a number of myths and shibboleths about television viewing have been destroyed, this should also engender some meekness when pronouncing on its effects.

10.5 Summary

Little is firmly understood about the social or individual effects of television. Even advertising, widely thought to be supremely effective, is seldom very powerful and is mostly defensive in its role.

IBA Reports

(Most of the results summarised in this book are based on these reports, prepared for the Independent Broadcasting Authority, London; some of the US findings come from working papers prepared for the John and Mary R. Markle Foundation, New York).

Qualitative Impressions of Audience Reach	Sept.	1967
The Factor-Analytic Search for Programme Types	Oct.	1967
Duplication of Viewing Between and Within Channels	Jan.	1968
The News in May	Feb.	1968
The Standard Programme Categories	April	1968
The Limits of the Inheritance Effect	June	1968
The News at Ten	June	1968
The Size of the Inheritance Effect	Aug.	1969
Higher Duplication	Aug.	1969
Sex and Age in Duplicated Viewing	Sept.	1969
Repeat-Viewing Week-by-Week	March	1970
TV Duplication of Viewing in June 1969	April	1970
Repeat-Viewing with Irregular Programming	Aug.	1970
Party Political Broadcasts	Nov.	1970
World Cup '70	March	1971
Housewives Viewing Intensity	Sept.	1971
Twenty Questions and Answers about Channel-Loyalty	Jan.	1972
The Audience of the News	Jan.	1972
Audience Build-up – Some Preliminary Findings	April	1972
Viewing Intensity and Programme Choice: Total TV	Dec.	1972
Viewing Intensity and Programme Choice: ITV	Dec.	1972
Repeat-Viewing and Audience Cumulation	Jan.	1973
Repeat-Viewing and the Audience Appreciation Index	April	1973
Audience Reaction to Different Episodes	July	1973
Viewing Intensity and Programme Choice (Supplement)	July	1973

Duplication of Viewing Amongst Regular Viewers   Aug.   1973
Mr. Trimble – The Viewing of a Pre-School Programme   Nov.   1973
Repeat-Viewing by Light Viewers – An Initial Analysis   Feb.   1974
Duplication of Viewing on One-Person Households   March   1974
Duplication of Viewing among Heavy and Light Viewers   April   1974
Set-On Data   Aug.   1974
Programme-Type Effects   Oct.   1974
55% Repeat-Viewing   Nov.   1974
Audience Retention Within a Programme   Nov.   1974

The Liking of Programme-Types   Jan.   1975
Repeat-Viewing among Light, Medium and Heavy Viewers   Jan.   1975
55% Repeat-Viewing   Jan.   1975
Updating to 1974   June   1975
Some Possible Programme-Type Effects   Oct.   1975
Some Correlates of the Appreciation Index   Nov.   1975
Duplication in AURA-Type Data   March   1976
Limitations of Factor Analysis   April   1976
Adult Education Programmes on ITV   May   1976
Repeat-Viewing on the AURA Panel   July   1976
Channel Reach   Sept.   1976
The News: Viewing Within a Day   Sept.   1976
Repeat-Viewing over Ten Weeks   Oct.   1976

The Development and Use of the Duplication of Viewing Law Jan.   1977
Audience Build-up over Ten Broadcasts   Feb.   1977
Programmes Viewers Like to Watch   Feb.   1977
Watching Television   May   1977
Duplication of Viewing in 1977   Aug.   1977
Repeat-Viewing of Some Series and Serials   Nov.   1977
Emmerdale Farm at Different Times of Day   Dec.   1977
Availability to View   Feb.   1978
Systematic Variations in the AI   March   1978
Repeat-Listening Across Weekdays   May   1978
The Welsh-Speaking Minority   Aug.   1978
The Regularity of Viewing Any Television   Sept.   1978
Glued to the Box   Nov.   1978
Radio Research over Five Days   Nov.   1978
Duplication between Radio Audiences   Dec.   1978

Weekly Reach and Demographics   March   1979
Audience Appreciation and Audience Size   April   1979

How Many People Watched Holocaust?   April   1979
Episodes and Programmes   June   1979
The Revenue Potential of Channel 4   Sept.   1979
Complementary Programming   Oct.   1979
Viewers' Average Appreciation Scores   Jan.   1980
Edward and Mrs Simpson   Jan.   1980
Station Switching   Feb.   1980
The Effort of Switching Channels   April   1980
Regional Programmes   Aug.   1980
Attitudes to Episodes and Programmes   Nov.   1980
Channel Reach in 1980   Dec.   1980

Duration of Listening   March   1981
Repeat-Listening Across Weekdays: Sheffield 1979   June   1981
Duplication between Radio Audiences: Sheffield 1979 July   1981
Interesting or Enjoyable?   Aug.   1981
Repeat-Showings   Sept.   1981
Alone or in Company   Sept.   1981
Radio Reach over 5 Days: Sheffield 1979   Oct.   1981
Some Insights from Canada   Nov.   1981
Audience Duplication and Programme Types   Feb.   1982
Repeat-Viewing: An Update   March   1982
Cable Expansion and Broadcasting Policy   May   1982
Brideshead and the Borgias   June   1982
Weekend Radio: Repeat-Listening   June   1982
How Much Does Television Cost?   July   1982
Weekend Radio: Audience Duplication   July   1982

Notes on Channel 4 Audiences   March   1983
Off-Peak Scrambled Television   June   1983
Repeat-Viewing: February 1983   June   1983
Multi-Channel Reach in a Day   July   1983
Viewing of Programme-Types   July   1983
Programme Audience Profiles by Channel   Nov.   1983
Jewel in the Crown: Compelling Viewing?   June   1984
Individual Viewing of Programme Types   July   1984
Amounts of Channel Viewing   Aug.   1984
Viewing of Programme-Type Audiences by Programme
   Audiences   Nov.   1984

References

Barnett, T. and Lougher (1971), 'Multi plus – a model for television planning and evaluation', *Admap*, January, 20–5.

Barwise, T.P. (1986), 'Repeat-viewing of prime time TV series', *Journal of Advertising Research* (in press).

Barwise, T.P. and Ehrenberg, A.S.C. (1984), 'The reach of TV channels', *Int. Journal of Research in Marketing*, I, 37–49.

Barwise, T.P. and Ehrenberg, A.S.C. (1985), 'Consumer beliefs and brand usage', *Journal of The Market Research Society*, 27, 81–93.

Barwise, T.P. and Ehrenberg, A.S.C. (1986), 'The liking and viewing of regular TV programs', *Journal of Consumer Research* (in press).

Barwise, T.P., Ehrenberg, A.S.C. and Goodhardt, G.J. (1982), 'Glued to the box?', *Journal of Communication*, Autumn, 22–9.

Belson, W.A. (1962), *Studies in Readership*, London: Business Publications.

Belson, W.A. (1967), *The impact of television*, London: Crosby-Lockwood.

Blumler, J.G. and McQuail, D. (1968), *Television in politics: its uses and influences*, London: Faber.

Bower, R.T. (1973), *Television and the public*, New York: Holt, Rinehart and Winston.

Bruno, A.V. (1973), 'The network factor in TV viewing', *Journal of Market Research*, 13, 5, 33–9.

Ehrenberg, A.S.C. (1966), 'Laws in marketing – a tailpiece', *Applied Statistics*, 15, 257–67. (Also in Ehrenberg, A.S.C. and Pyatt, F.G., eds (1971), '*Consumer behaviour*', London: Penguin Books.)

Ehrenberg, A.S.C. (1972), *Repeat-buying*, Amsterdam: North Holland; New York: Elsevier.

Ehrenberg, A.S.C. (1974), 'Repetitive advertising and the consumer', *Journal of Advertising Research*, 14, 2, 25–34.

Ehrenberg, A.S.C. (1982), *A primer in data reduction*, London and New York: Wiley.

Ehrenberg, A.S.C., Goodhardt, G.J. and Barwise, T.P. (1986), 'Brand popularity and brand loyalty: the double jeopardy effect', *Journal of Marketing* (in press).

Ehrenberg, A.S.C. and Wakshlag, S. (1986), 'Repeat-viewing in the Boston panel', *Proceedings of the ARF Electronic Media and Research Technologies Workshop*, New York.

Fawley, A. and Fairclough, E. (1972), 'The prediction of coverage and four-plus for television schedules', *Admap* February, 74–6.

Frank, R. and Greenberg, M. (1972), *The public's use of television*, Los Angeles: Sage Publications.

Gensch, D.E. and Ranganathan, B. (1974), 'Evaluation of television program content for the purpose of promotional segmentation', *Journal of Marketing Research*, 11, 390–8.

Goodhardt, G.J. and Chatfield, C. (1973), 'The gamma distribution in consumer purchasing', *Nature*, 244, 316.

Goodhardt, G.J., Ehrenberg, A.S.C. and Chatfield, C. (1985), 'The Dirichlet: a comprehensive model of buyer behaviour', *Journal of Royal Statistical Society (A)*, 147, 621–55.

Halloran, J.D., ed. (1970), *The effects of television*, London: Panther Books.

Halloran, J.D., Brown, R.L. and Chaney, D.C. (1970), *Television and delinquency*, Leicester University Press.

Headen, R.S., Klompmaker, J.E. and Rust, R.J. (1979), 'Duplication-of-viewing law and television media schedule evaluation', *Journal of Marketing Research*, 16, 333–40.

Himmelweit, H.T., Oppenheim, A.N. and Vince, P. (1958), *Television and the child*, London: Oxford University Press.

Howitt, D. and Cumberbatch G. (1975), *Mass media violence*, London: Paul Elek.

Hulks, R. and Thomas, S.G. (1973), 'Preface: A simple model for the prediction of television coverage and frequency distribution', *Admap*, December 349–53.

Hyett, G.P. (1958), *The measurement of readership*, Seminar at the London School of Economics.

ITA (1970) *Religion in Britain and Northern Ireland: a survey of popular attitudes*, London: ITA.

Johnson, D. and Peate, J. (1966), 'The estimation of television viewing frequency', *Admap*, July/August, 394–6.

Kirsch, A.D. and Banks, S. (1962), 'Program types refined by factor analysis'. *Journal of Advertising Research*, 2, 3, 29–32.

Klapper, J.T. (1960), *The effects of mass communication*, New York: Free Press.

Liebert, R.M. Neale, J.M. and Davidson, E.S. (1973), *The early window*, Oxford and New York: Pergamon.

Marplan (1965), *Report on a study of television and the managerial and professional classes*, London: Marplan.

McPhee, W.N. (1963), *Formal theories of mass behaviour*, New York: Free Press.

Metheringham, R. (1964), 'Measuring the net cumulative coverage of a print campaign', *Journal of Advertising Research*. 4, 23–8.

Pilkington Committee (1960), *Report of the Committee on Broadcasting*, London: HMSO.

Pragnall, A. (1985), *Television in Europe*, Manchester: European Institute for the Media.

Rubenstein, E.A., Comstock, G.A. and Murray, J.P., eds (1972), *TV and Social Behaviour*, Washington D.C.: Government Printing Office.

Schuchman, A. (1968), 'Are there laws of consumer behaviour?' *Journal of Advertising Research*, 8, 19–27.

Segnit, S. and Broadbent, S. (1973), 'Life-Style research', *European Research*, 1, 6–13.

Steiner, G.A. (1963), *The people look at television*, New York: Alfred A. Knopf.

Swanson, C.E. (1967), 'The frequency structure of television and magazines, *Journal of Advertising Research*, 7, 2, 8–14.

US Department of Health, Education and Welfare (1972), *Television and growing-up: the impact of television violence*, Washington D.C.: Government Printing Office.

Williams, R. (1974), *Television*, London: Fontana.

Index

Indexer's note: specific television programmes referred to in the text are listed together under 'Programme titles'.

Adventure programmes/series 100, 107, 112
Advertisements/advertising 1–5, 63f, 88, 121–3 (effects)
Afternoon viewing *see* Off-peak viewing
American Broadcasting Company (ABC) 3, 13f, 77–9, 81–4
American Research Bureau (Arbitron) 6, 79
Appreciation Index 7, Chapter 8, 105
Arts and TV 119
Arts documentaries/programmes 11f, 119
Attitudinal measures 7
Audience:
  appeal 15, 106–9
  appreciation 7, Chapter 8, 114, 116
  appreciation panels 7
  characterisation/make-up 8
  cumulation 62–7
  duplication 9–13, 21f, 35, 77–86 (US), 115f
    BBC1 and BBC2 29f
    BBC1 and ITV 25f, 31
    BBC2 and ITV 29f
    on Channel 4 32f
    on non-consecutive days 24
    of same programme *see* Repeat-viewing
  flow Chapter 2, 19 (across channels, Chapter 7 (US))
  measurement 6
  overlap *see* Audience duplication
  size *see* Ratings
Audits of Great Britain (AGB) 7, 68
Availability factor 11, 28, 50, 55, 58, 82, 115
Average frequency 63–5

Banks, S. 41
Barnett, T. 68
Barwise, T.P. 56, 77
BBC 1 2, 5, 71 (hours viewed)
BBC 2 2, 5, 38 (high duplication), 72 (hours viewed), 117
Belson, W.A. 120f
Beta Binomial Distribution (BBD) Model 65–7
Birmingham, Alabama 77
Blumler, J.G. 120
Bower, R.T. 116
British Broadcasting Corporation (BBC) 2, 4, 6, 8, 40, 76, 87, 118
Broadbent, S.R. 100
Brown, R.L. 120
Bruno, A.V. 45
Business programmes 119

Cable television 3, 117
Chaney, D.C. 120
Channel loyalty Chapter 3, 19 (defined), 35, 40, 42, 45, 81, 83, 86, 115
  to BBC1 27f
  to BBC2 27–9
  to ITV 20–5
Channels:
  choice of 2f, 70–3
  number of 2, 86
  public service 117
Channel switching 25–7, 81, 86
Channel 4 2, 57, 60f, 70–3, 76
Chatfield, C. 68
Children's programmes 5, 10, 113
Columbia Broadcasting System (CBS) 3, 45, 77–9, 83–5
Community stations 2

Cumberbatch, G. 120
Current affairs programmes 14, 100–4,
    108, 111, 113

DBS (Direct broadcasting by satellite) 3,
    117
Demographic factors 35, 55
Double jeopardy 55f
Drama programmes 14, 54, 56, 121
Duplication coefficient (*k*) 10f, 17, 20,
    26, 31–3, 81–3
Duplication of viewing law 10–3 (de-
    fined), 16–8, 22–8 Chapter 4, 32f
    (time factor), 46 (news bulletins),
    77, 81–6 (US), 115, 120
  deviation/exceptions 11, 15, 24, 27, 35
  empirical basis 11
  theoretical explanation 115

Early evening:
  effect 35f
  viewing *see* Off-peak viewing
Education and TV 120
Educational programmes 5
Effects of television 120–4
Ehrenberg, A.S.C. 15, 45, 56, 77, 122f
Election broadcasts 118
'Enforced' viewing 110f, 113
Exposure distributions 62–7

Factor analysis 45
Fairclough, E. 68
Fawley, A. 68
Federal Communications Commission
    (FCC) 2f
Films 3, 14, 56
Frank, R. 8
Frequency of viewing 93–8 (and AI),
    113, 115

General entertainment programmes
    see Light entertainment programmes
Gensch, D.H. 45
Germany 2
Goodhardt, G.J. 68
Government monopoly 1f
Greenberg, M. 8

Halloran, J.D. 120
Headen, R.S. 45
Heavy viewers 71–3, 95, 99, 102
Himmelweit, H.T. 120

Howitt, D. 120
Hulks, R. 68
Hyett, G.P. 68

Independent Broadcasting Authority
    (IBA) 2, 4, 7, 87, 103, 118
Independent channels/stations 2f, 77, 83,
    86
Independent Television Authority (ITA)
    120, *see also* IBA
Inheritance effect 11, 35, 40–2, 77, 82–6
    (US)
Intensity of viewing *see* Viewing intensity
Intentions to buy 88
Irregular viewing 112f, 114
ITV 2, 71 (hours viewed), 87

Johnson, D. 68
Joint Industry Committee for Television
    Advertising Research (JICTAR) 92

Kirsch, A.D. 41
Klapper, J.T. 120

Lapsed viewers 57
Las Vegas 77
Late-evening viewing *see* Off-peak
    viewing
Lead in effects 41f, 84–6
Lead out effects 42, 84–6
Leo Burnett Company 100, 103f
Leo Burnett Life Style Research Study
    100, 110
Liebert, R.M. 120
Light entertainment programmes 100,
    102–12
Light viewers 71, 75f, 95
'Liking' Chapter 9
Lougher, J. 68

Marplan 116
McPhee, W.N. 55f
McQuail, D. 120
Media planning 64
Metheringham, R. 68
Minority interests/programmes 117–9

National Association of Broadcasters 3
National Broadcasting Company (NBC)
    3, 77–9
National Commission on the Causes and
    Prevention of Violence 123

Networks 3, 6
  national 3, 19, 77, 81, 86
  regional 3, 77, 86
News programmes 5, 13, 45–9
New York, 13, 77, 80–3
Nielsen, A.C., Company 6
Non-repeat viewing 58
Non-stationarity 56f, 66f

Off-peak viewers/viewing 35–8, 82 (US),
  115
One-person households 45
Oppenheim, A.N. 120

Peate, J. 68
Persuasion 122
Pilkington Committee 88, 90
Political braodcasts 118
Politics and TV 120
Population 15 (defined)
Pragnall, A. 2
Prediction 64
Programme:
  appeal *see* Audience appeal
  categories/types 5 (defined), 42–5, 54,
    90–2, 97f Chapter 9, 114
  choice (actual behaviour) 74f
  clusters Chapter 9, 116
  companies 2
  content 4, 9, 13, 16, 42, 67, 114f
  effects 35, 42–5, *see also* Effects of TV
  material 124
  type preferences 42–5, 50, Chapter 9,
    110–3 (*vs.* viewer behaviour)
Programmes:
  available 4f, 117
  'cult' 104
  low-rating 76
  thirty-minute 16
Programme titles:
  ABC News 13
  Adventurers, The 20–2, 37
  Aquarius 107f
  Bless This House 22, 37
  Boxing 105
  Brady's Bunch 82
  Brideshead Revisited 62–4, 67
  Cannon 120
  Cinema 12
  Contours of Genius 12
  Coronation Street 96f, 107
  Eyewitness News 13

Family at War 106–7
Flintstones, The 5
4.30 Movie, The 13
Galloping Gourmet, The 108
Golden Shot, The 105
Hawaii 5-0 74, 100
Ironside 120
Jason King 107
Jewel in the Crown 54
Lassie 4
Line-Up 107
Lost in Space 36
Love American Style 13
Lucy Show, The 5, 78
Man at the Top 93, 95
Mary Tyler Moore Show, The 117
Match of the Day 105, 119
Midweek 14
Mind of J.G Reeder, The 20f, 42
Monday Play, The 14
Monty Python 108
Morecambe and Wise 105
News at Ten 9, 21, 37, 42, 49f
Opportunity Knocks 21, 105
Owen, M.D. 108
Panorama 14, 107, 110
Persuaders, The 100
Peyton Place 107
Public Eye 107
Rugby Special 101, 105
Saint, The 22, 37, 42
$ Six Million Man, The 82
Startrek 5, 14
Talk-Back 108
This is Your Life 24, 63–5, 107
This Week – The Arts 12
Today 37
Tuesday Film, The 14
World in Action 21, 42
World of Sport, 101, 105
Wrestling 105
Z-Cars 51–3, 56
Programming Chapter 1:
  balance/variety 2–5, 118
  competitive 34, 49, 88
  complementary 29f, 49, 117
  irregular 114
  policy 35, 45–9, 117f
  repetitive 4f, 13
Proportionality factor *see* Duplication
  coefficient
Public affairs programmes *see* Current
  affairs programmes

Quiz programmes *see* Games and quiz
    programmes

Ranganathan, B. 45
Ratings 9–18 (defined), 22, 51, 76, 79–81
    (and AI), 87–92, 98, 112 (*re* viewing
    claims), 114f (and repeat-viewing),
    53–60, 80
Reach 63–9
Religion and TV 120
Religious programmes 120
Repeat-viewing 8, 13f (defined),
    Chapter 5, 58 (non-consecutive epi-
    sodes), 76–80 (same week), 86, 87,
    93–7 (and AI), 114f, 118, 120
  individuals 93–7
  national networks 77–9
Re-runs 78
Rubenstein, E.A. 120

Sampling errors 15, 80
Schuchman, A. 56
Seasonal factors 39
Segnit, S. 100
Selective viewing 75f, 120
Serials 4f (defined), 14, 54, 114
Series 4f (defined), 14, 51, 54, 115, 117
Situation comedy programmes 42
Social research 120
Sports programmes 100f, 105, 108–12
Stationarity 65 (defined), 79
Statistical:
  methodology 15
  significance 15
Steiner, G.A. 116f
Strip-programming 5 (defined), 77–9, 82,
    86
Swanson, C.E. 45

Television Audience Measurement Ltd
    (TAM) 68 (reach and frequency
    guide)
Television, background 1–4

Television sets, number of 1
Thomas, S.G. 68
Time-bands 16
Times, The (London) 9
Transmissions, start of 1(UK), 2(US)
Trivialisation 121

Ultra-high frequency (UHF) 2
United States 2f, 13f, Chapter 7
US Department of Health, Education
    and Welfare 120

VCR (Video Cassette Recorder) 3, 61,
    117
Very high frequency (VHF) 2
Viewer:
  diaries 6
  panels 6
  satisfaction *see* Audience appreciation
Viewers:
  habitual 11, 120
  repeat, new and lapsed 94f
Viewing:
  behaviour, US 13f, Chapter 7
  compulsive 120
  habits 97
  intensity Chapter 6:
    other channels 57f, 61, 73
    hours 8f, 70–3
Vince, P. 120
Violence, and TV 120f

Wakshlag, S. 77
WABC 78, 82
WCBS 78f, 82
Week-end viewing 35, 38f, 55
Westerns 42, 118
Williams, R. 4f
WNBC 78, 82
WNET 77
WNEW 77, 79f
WOR 77, 79
WPIX 77, 79

DATE DUE

DEC 0[...]

DEC 2[...]

DEC 2[...]

MAR 2 0 1987

APR 11 1997

APR 16 RECD

AUG 07 1998

JUN 1[...] 1999

MAY 0[...] RET'D

OCT 2[...]

NOV 2 3 1993

NOV 17 RET'D

MAR 2 0 1987

SEP 0 4 RECD

NOV 02 2000 RECD

DE[...]

FEB 0 4 1994

FEB 2[...] RECD

MAY 0 [...] 1994

NOV 2 5 RECD

MAR 2 7 2000

MAR 2 8 RET'D

MAY 0 6 1994

NOV 2 8 1994

NOV 0 6 1995

MAY 0 5 [...]

MAY [...] RECD

NOV [...] RECD

DEC 0 1 1995

NOV 2 7 RET'D

MAR 2 0 1980

RET'D

RET'D

APR 20 1998

AUG 0

AUG 1 1 '8[...]

RET'D

2[...]

AUG 1 1 1989

JUL 1 2[...]

AUG 1 1 1989

DEMCO 38-297

DEMCO 38-297

A113 0978519 1

BOWLING GREEN STATE UNIVERSITY
DISCARDED
LIBRARY

HE 8700.66 .G7 T44 1987

The Television audience